Paton,
Drummond Macdonald.

Man and mouse

DATE			

MAN
AND MOUSE

MAN AND MOUSE

ANIMALS IN MEDICAL RESEARCH

WILLIAM PATON

Oxford New York
OXFORD UNIVERSITY PRESS
1984

Oxford University Press, Walton Street, Oxford OX2 6DP
London New York Toronto
Delhi Bombay Calcutta Madras Karachi
Kuala Lumpur Singapore Hong Kong Tokyo
Nairobi Dar es Salaam Cape Town
Melbourne Auckland
and associated companies in
Beirut Berlin Ibadan Mexico City Nicosia

Oxford is a trade mark of Oxford University Press

First published 1984 as an Oxford University Press paperback
and simultaneously in a hardback edition

British Library Cataloguing in Publication Data
Paton, Sir William
Man and mouse—(Oxford paperbacks)
1. Diseases—Animal models
I. Title
619 RB125
ISBN 0–19–217734–6
ISBN 0–19–286043–7 Pbk

Set by Hope Services, Abingdon
Printed in Great Britain by the
Guernsey Press Co. Ltd.,
Guernsey, Channel Islands

SSH

FOREWORD

SIR ANDREW HUXLEY

The strength of feeling against inflicting pain on animals is one of the reasons why Britain took little part in the rapid advance in physiology—the science of the workings of the body—that took place in the early and middle decades of the last century. This development, chiefly in France and Germany, depended heavily on experiments involving major surgery on unanaesthetized animals, and accounts of some of the experiments are horrifying to read. Experimental physiology did not get going in Britain until around 1870, by which time the use of anaesthetics had become general. Nevertheless, there was much public debate at that time against experimentation on living animals, and this led to the appointment of a Royal Commission in 1875 and finally to the passing of the Cruelty to Animals Act of 1876, which is still in force and is the basis on which the Home Office regulates animal experimentation.

Since those days, agitation has continued but has fluctuated in intensity. Articles by the secretary of the National Anti-vivisection Society in the *Daily News* led, in 1903, to a libel action in which Dr William Bayliss (later Professor of General Physiology) of University College London was awarded very substantial damages. Agitation was again strong in the 1930s, and we are now in the midst of another wave, in which terrorist groups break into laboratories and use letter-bombs to intimidate those who dare to remind the public of the advances in curative medicine and in public health that have followed from experiments on living animals.

As in many matters of public concern, it has become practically impossible for the man in the street to form a sensible judgement on this issue. Scientists are accused of inflicting severe and unnecessary pain and it is claimed that equivalent results could be obtained by 'alternative methods'; these assertions are denied; and there is no way in which a layman can discover where the truth lies. The Press, radio, and television concentrate on sensational accusations: the *Daily Star* recently ran a campaign against the Institute of Animal Physiology at Babraham, near Cambridge, with headlines almost every day using words like 'Horror Farm' and 'Torture', but when the Institute was inspected in 1983 by the House of Commons Select Committee on Agriculture their report said: 'We are more than pleased to record our great relief at the very high quality of animal care that we found at Babraham where important work on animal physiology is carried out in a most humane manner.' This conclusion was not reported in the national Press, but even if it had been, how would the layman know whom to believe?

This book by Sir William Paton, Professor of Pharmacology in the University of Oxford, sets out, with great success, to give a background of information about the way in which medical science progresses, so as to help the reader to judge where the line should be drawn when considering experiments on living animals. It is particularly timely, as the Government published a White Paper in May 1983 with its proposals for new legislation to replace the Act of 1876, and has said that it will bring in a Bill 'as soon as Parliamentary time permits'. This book will do much to prepare the ground for an informed debate on this difficult, important, and controversial subject.

CONTENTS

LIST OF FIGURES

LIST OF TABLES

ACKNOWLEDGEMENTS

This book owes much to many discussions and debates in recent years. I am particularly grateful to my wife for help in reducing jargon, to Sir Richard Doll, Professor Henry Harris, and Dr P. O. Williams for detailed comment, to my secretary Mrs Nesta Dean for dealing with a series of palimpsests, to Miss Nicola Bion for searching editorial criticism, and to Mr Patrick Hunter for indexing skill.

1

INTRODUCTION

The debate about animal experiment is an old one—at least 150 years old. One of the most interesting early documents on the subject is a paper written in 1831 by Marshall Hall, a distinguished neurologist and pioneer of physiology. In it he puts forward his own proposals for the regulation of physiological experiments, which were just beginning to throw a new flood of light on our knowledge of the working of the animal body.

> The first principle to be laid down for the prosecution of physiology is this: we should never have recourse to experiment, in cases in which observation can afford us the information required . . .
>
> As a second principle . . . it must be assumed that no experiment should be performed without a distinct and definite object, and without the persuasion, after the maturest consideration, that that object will be attained by that experiment, in the form of a real and uncomplicated result . . .
>
> It must be admitted, as a third principle . . . that we should not needlessly repeat experiments which have already been performed by physiologists of reputation. If a doubt respecting their accuracy, or the accuracy of the deductions drawn from them, arise, it then, indeed, becomes highly important that they should be corrected or confirmed by repetition. This principle implies the necessity of a due knowledge of what has been done by preceding physiologists . . .
>
> . . . it must next be received as an axiom, or fourth principle, that a given experiment should be instituted with the least possible infliction of suffering . . .
>
> Lastly, it should be received as a fifth principle, that every physiological experiment should be performed under such circumstances as will secure a due observation and attestation of its results, and so obviate, as much as possible, the necessity for its repetition . . .
>
> In order fully to accomplish these objects, it would be desirable to form a society for physiological research. Each member should engage to assist the others. It should be competent to any member to propose a series of experiments, its modes, its objects. These should be first fully discussed—purged from all sources of complication, prejudice, or error—or rejected. If it be determined that such series of experiments be neither unnecessary nor useless . . . they should then be performed, repeated if necessary, and duly attested. Lastly, such experiments, with the deductions which may flow

from them, may then be published with the inestimable advantage of authenticity.

Pursued in this manner, the science of physiology will be rescued from the charges of uncertainty and cruelty, and will be regarded by all men, at once as an important and essential branch of knowledge and scientific research.[1]

An old debate in a new context

As one reads this, many of the issues discussed today can be recognized: consideration of alternatives, defined objectives, avoidance of repetition, minimum suffering, proper record, open criticism. These remain the objectives today. Are we then merely raking over the embers of an old fire, that sometimes dies down, sometimes flares up, over which men will never reach agreement? Or is there something new to be said? This book will argue that both of these statements are true. The question of animal experiment depends in the end on the view man takes of himself and of the animal and inanimate world around him: it focuses on his choice between the avoidance of suffering now and the avoidance of suffering and remediable ignorance in the future. I do not believe there is any set of rules, any 'algorithm' suitable for a computer, which allows an unequivocal answer to the question, but rather that it forms part of the familiar pattern of moral decisions of everyday life about which men have differed, do differ, and will differ. But at the same time the debate today is in a different context from that in the past, and it is worth looking at some of the reasons for this.

One obvious fact, apparent to anyone who considers animal and human life in past centuries, or even decades, is a striking *rise in standards* of physical and mental well-being. Even as late as the 1930s, a schoolboy could have lost a companion from tuberculosis, mastoid infection, diphtheria, or scarlet fever, and might play with a friend crippled by polio. Deformity, pain, and disability were familiar experiences. For the routine pains of sinusitis, colic, peptic ulcer, or 'rheumatism', laudanum (or alcohol) was available for those who could afford it in the last century, and aspirin in this; but a great deal of suffering was simply accepted as part of life. Cruelty for its own sake seems always to have been reprobated; but parents who had lost most of their family in infancy, and learnt to speak coolly of it, could readily be equally cool about the suffering of animals. Today, however, in comparison, most people in the industrialized countries are healthy, straight-limbed, and pain-free;

childhood ailments no longer threaten the bloom of childhood and adolescence—so rapidly lost in the past—and we can expect with some confidence a life-span of threescore years and ten in which, if disease intervenes, effective palliation at least will be available. The care of animals has improved in parallel, spilling over quite naturally, one could suppose, from man to his domestic companions and thence to animals generally. The same therapy, the same local or general anaesthetics, the same surgery, are used to abate their suffering as are used in man. Just as with man, for whom 'mental distress' may now be the cause of substantial compensation in the courts, so the animal-lover broadens his concern from its focus on pain to include 'distress' or 'suffering' in a more general sense. It is important to recognize that the debate takes place in a context where standards have risen greatly, and must be expected to go on rising.

A second factor, bringing animal experiment more to the fore, is its *growth in scale*. The extent of domestication of animals (with all the scope it offers for ruthless exploitation or unintended infliction of suffering), the use of animals for food and other purposes, their use in hunting, and the severity of animal life in the wild, have not, one might guess, changed greatly; but the last hundred years have seen a very considerable growth in animal experiment, now levelling off or declining, in the UK at least. The discovery, essentially in the past century, of how animal experiment (amongst other methods) can lead to great advances in knowledge and to equally great benefits for men and animals, has inevitably led to its *adoption by industry*. This is vitally important, for it makes available to the many what would otherwise be available only to the few. As a result most animal experiment, estimated by the animals used, is now carried out by industry. At the same time, the rise in standards of health, further catalysed by consumer pressure, has reflected itself in the *growth of regulatory bodies* concerned to maintain and improve these standards. To do this such bodies themselves lay down animal tests that must be done before medicaments, foods, and chemicals used in the environment, indeed *any* chemical to which workers may be exposed, are allowed to be made generally available by industry. The involvement of industry and hence of technological change can bring more general criticisms into the argument—of capitalism, of interference with nature, of modern medicine, and indeed of scientific investigation itself.

One cannot hope, therefore, to achieve any finality of decision.

Moral debate is inevitable, and one must also recognize that the material of the debate is changing. Standards will go on rising, both in the search for further improvement in human and animal well-being, and in the attempts to ensure safety in the way those improvements are brought about. Techniques will improve to meet these demands. Continued adjustment will be obligatory.

The nature of the argument

There is another change to notice. A century ago, the two sides in the argument were fairly well balanced: there was not a large body of animal experiment; so the antivivisectionist could well familiarize himself with and genuinely understand a good deal of it, while the experimenter had time to answer the charges. Scientific journals such as *Nature* and the *Lancet* were more widely read. Literary journals such as the *Nineteenth Century* published scientific material. Laymen could, if not take part, at least align themselves with either side; and one finds bishops, lawyers, MPs, the aristocracy, and literary men, lending their names freely one way or the other. So even if, from the start, the debate was polarized at the extremes, and charges of cruelty, ineffectiveness, and arrogance were liberally exchanged with those of misrepresentation, obscurantism, and a deeper cruelty, yet both the debate, and its materials, were reasonably open.[2]

But the situation changed. Scientific knowledge expanded enormously; it became more demanding of the investigator, and harder for the layman to understand. Biological or medical science, like other branches of science, became segregated as full-time professional careers. Science and medicine became institutionalized, so that an 'establishment' began to grow. At the same time, as the demands on the scientist grew, explanation by scientists themselves for laymen of what they were doing diminished: this was partly as a result of the growth of learned societies, which took the place of the British Association as the scientists' professional forum. Descriptions for the layman of scientific work by the scientist began to go by the board, apart from the activities of interpreters, 'popularizers of modern science' (the very phrase expresses the distance of actual scientific practice from the public's general understanding). Indeed, a good many medical scientists seemed to begin to feel that it was better to let the achievements of science and medicine speak for themselves, and, in modern jargon, to 'adopt a low profile'. So common ground for the materials for debate no longer exists. On the one hand there are antivivisectionists and more or less radical animal-welfare groups: most of their arguments are non-technical and easily accessible to any thinking person, they are well endowed by benefactions from sympathizers, and most Members of Parliament receive a regular mail from them. On the other hand we see the large body of scientific and medical practice—together with their appropriate institutions—and a range of major industries, both pharmaceutical and of other types, involved with animal experiment. These command considerable resources yet make relatively little public contribution to the general debate: this is left to one or two small groups speaking for the experimenter. The public is thus very poorly informed about the great body of current and past scientific work, apart from accounts in the press of 'breakthroughs' from time to time. One cannot be confident that many laymen could even give a reasonable account of the working of their own heart and circulation, or of what happens to the food they eat.

The situation has been further exacerbated by the scientific inadequacy and prejudice of some animal-welfare comment. One may read, for instance, that a leading animal campaigner believes that 'anyone with an ounce of medical knowledge would have known that a volatile anaesthetic doesn't get into the bloodstream',[3] although the fact is that it is only through the bloodstream that it passes from lung to brain; or that it is 'almost without doubt' that the only experiments in the medical sphere are those involved in cancer research, diagnostic procedures, and drug research, and that 'high-grade medical research work—the sort of work being done not in commercial laboratories but in universities—probably accounts for only a few thousand animals each year'.[4] Or one may find in a distinguished daily paper the same picture of a monkey apparently smoking a cigarette inserted twice into articles on animal experiment in Britain, only to learn later that it is a Russian picture, with no explanation or reference, distributed by the Tass News Agency.[5] The scientist tends to react to all these things with contempt, which in turn generates heated imputations of arrogance, and comparisons of animal experiment with racism or Nazism, matching the comparisons made with slavery in the last century.

Another unhappy development has been the rise of physical violence by an extreme wing of the animal-welfare movement. This inevitably reduces the willingness of individual scientists to come forward to explain their position: too often this has led to personal or family harassment or other damage, or to harassment or physical damage to the institutions they work for.

A last, potentially more beneficial event has been the entry of moral philosophers into the debate, partly as a result of the professional philosophical interest in the questions presented by man's experience of pain.

The result of all these forces is a fairly voluminous animal-welfare and antivivisectionist literature, including some recent substantial books on animal rights and animal welfare, but a distinctly thin literature on the animal experimenter's position. There is one study of 'alternative methods', but no contemporary book on the general biomedical arguments other than Sir Leonard Rogers's small, trenchant, but not generally known volume, *The Truth about Vivisection*,[6] published in 1937, and Lepage's *Conquest*,[7] published in 1960.

The present book is an attempt to go some way in redressing the balance. The pattern of it is in part dictated by the views mentioned above. Because the nature of biomedical work and its achievements has been poorly explained, some attempt will be made to do this. Dogmatic statements about 'great advances' and 'breakthroughs' have seemed to me, in public discussions, almost totally useless. But particular instances, with an adequate explanation of the specific scientific issues involved, of the research opportunity or requirement that prompted the work, of the procedures used (and the reasons for their choice), and of the outcome in new knowledge and benefit, seem consistently to arouse interest and respect, even in those who remain opposed.

It will also be necessary to spend some time on general arguments, although the approach generally will be that there are no 'knock-down' arguments, that moral philosophy may clarify but does not prescribe, and that in the end it is a matter for collective moral decision. The layout is not strictly logical, but is that which has seemed to arise naturally in argument, to a large extent centring round particular recurrent questions. A number of special issues will be taken up in separate sections.

Finally, although there can be deep feelings about the issues, it seems clear from the past that too vigorous expression of personal attitudes merely deepens existing entrenched positions. Any writer is bound to have his own views, and these cannot be disguised. The view taken here, however, is that an argument gains no special validity because a particular individual holds it (or opposes it), nor because of the strength of his opinion. The objective is to lay out as much of the argument and evidence as possible in a generally accessible manner, so that the reader can judge for himself.

2

WHAT IS INCLUDED UNDER THE WORD 'ANIMAL'?

For there is in this Universe a Staire, or manifest Scales of creatures, rising not disorderly, or in confusion, but with a comely method and proportion: between creatures of meer existence, and things of life there is a large disproportion of nature; between plants and animals or creatures of sense, a wider difference; between them and man a far greater; and if the proportion hold on between men and Angels, there should be yet a greater.

Sir Thomas Browne, *Religio Medici* (1643), section 33

Our first choice of meaning might well be like that made by George Orwell in *Animal Farm*:[1] Boxer the horse, dogs, pigs, sheep, cows, goats, donkeys, the cats, ducks, geese, pigeons, turkeys, the ravens—even rats (voted in, by the others, with dog and cat dissenting). A restriction to non-human vertebrates, particularly the warm-blooded, certainly corresponds to one pattern of use. But the term can be used much more widely. The discovery of the microscope 300 years ago opened up a new world, and Leeuwenhoek used the term 'little animals' for creatures as small as protozoa and microbes.[2] Albert Schweitzer, with his reverence for life, records that he wondered, as he looked down his microscope, if he had been justified in killing the micro-organisms displayed there.[3] What are we to say of the frog, the fish, the bee, the ant, the wasp, or the lobster? Or does 'animal' simply mean anything not a vegetable or a mineral, and is that the range of creatures that we wish to provide for?

Once we broaden our view in this way, it is obvious that our concern is not with the semantics of the word 'animal', but with a deeper issue. As we survey the whole world around us—and we can consider nothing less than that, if we think of the range of scientific enquiry and practical use—can we see lines of demarcation? In past centuries such a panorama was implied in the idea of a Chain of Being, *Scala Naturae*, a Ladder of Perfection, stretching from the mineral world through lower forms of life up to man, and then to purely spiritual existences:[4] 'There is in this Universe a Staire', wrote Sir Thomas Browne, 'rising not disorderly, or in confusion, but with a comely method and proportion'. We can envisage such a panorama today, although that 'stair' takes a different aspect. It is no longer seen as a smooth development from one creature to another. Today we recognize that each species itself stands as the latest expression of one of a vast number of evolutionary branches. Evolutionary relationships might, indeed, help us in drawing lines of demarcation; but we also know that the same function—say flying, swimming, or seeing—can be achieved by creatures following totally different evolutionary pathways. So we cannot in general rely on finding some evolutionary branch-point that will decisively mark the point of appearance of those characteristics in which we are interested.

Indeed, what is it in which we are interested? One demarcation at issue is an upper bound, namely whether man is to be distinguished from animals and the rest of creation. But a lower bound is needed too. We do not propose to legislate for cruelty to clay or gravel, nor, one imagines, for amoebae or cabbages. So where do we place the line where our particular care about experimental work stops? In question, too, are the criteria by which we identify such demarcations. There are many proposals; and both they, and the difficulties they bring, are conveniently illustrated by looking at some of the links in the 'chain of being'. Even in the inorganic world we note that some things, like crystals, grow. Moving on to the organic world, we find that viruses, sometimes consisting of only a few different (though large) molecules, can reproduce in a suitable host cell. Bacteria can reproduce, both sexually and asexually, swim, and emit noxious substances. In the vegetable kingdom, flowers move, turning to the sun. The clematis curls its prehensile tentacles round the twigs or wires it climbs up. The sensitive plant, *Mimosa pudica*, responds to sound, folding its leaves at the noise of a handclap. Insectivorous plants such as the sundew (*Drosera*), as Charles Darwin showed,[5] are exquisitely sensitive to chemicals—one thousand millionth of a gram of ammonium acetate being enough to make one of its hairs bend. Moving on to the invertebrates we notice that even so simple a creature as the jellyfish has a nervous system and can respond to stimuli. If we watch an ant or a bee, or read about the powers of an octopus, it is hard not to admit the words 'learning' or 'purpose' in some sense in describing their behaviour. The leech and the snail[6] contain receptors for pain-killing drugs, and fish are rich in the opiate-like substances (peptides) similar to those in our own brains. There is no need to elaborate on the range of behaviour possible for cold-blooded or warm-blooded vertebrates, ranging from the frog through birds and primates to man. I do not think anyone has found a sharp, rigorous point of division in searching for our lower bound, between 'animal' and 'non-animal', whether one considers growth, movement, reproduction, self-protection, response to stimuli, sensitivity, learning, apparent purposiveness, or relationship to pain mechanisms familiar in man.

Likewise for our upper bound. We do not need to *prove* that animals other than man can suffer, or solve problems, or experience consciousness or fear. It is enough to note that no reason exists why they should not share these capacities, and that as our knowledge of animal and primate life develops, the more frequently vestiges of behaviour are recognized that can plausibly be compared to the human experience of pain, apprehension, sorrow, frustration, guilt, altruism, laughter, symbolic communication, thought, or invention. The behaviourists can, of course, formulate models for such behaviour—after all, machines can be designed to solve problems; and it will probably never be possible to exclude the possibility that behaviour, however complicated, is ultimately describable as the sort of machine that the French philosopher Descartes conceived an animal to be 300 years ago. But this seems equally unlikely ever to be provable. So we need to think of the panorama of creation with its millions of species—each species itself blurred by varieties and normal variability within the species—as presenting a continuous spectrum, with man and his characteristic qualities included.

To find such continuity and lack of dividing lines is a very common human experience. For instance, there is a continuous gradation between, say, lightness and darkness, with no divisions. Yet in practice we can distinguish the two ideas and use the words 'light' and 'dark' effectively, as in specifying the darkness of a photographic dark room, or the lightness of a workplace or street at night. More painfully, perhaps, we set levels in the continuum of income at which taxation changes, a level of blood alcohol that is liable to prosecution, levels of scholastic achievement to gain particular qualifications, and levels of speed, distance, or height that athletes must meet to qualify for competition. A great deal of human life consists in solving the problem of drawing a line at some point on a continuous scale, in order to allow some type of decision to be practicable.

How this is done would be a major study in itself. But a simple example, that of the division of the twenty-four hours into 'night' and 'day', points to two obvious processes. The first is to decide that the divisions should be made around sunset and sunrise, because these are the times at which the rate of change from one phase to the other is fastest; this narrows the zone of potential ambiguity. Then the final detailed definition, where it is necessary, can be made to depend ultimately on a general agreement on some acceptable and practicable defining procedure—for instance, using astronomical data, establishing time zones, or taking account of industrial needs by introducing 'daylight saving'.

The upper bound: man–animal

So we return to consider our upper and lower bounds for the 'animal' world, admitting continuity, but asking now whether there are practical, operational, generally accepted grounds of demarcation. As to the upper bound, I believe there is in fact a general consensus recognizing a major operational difference between man and other vertebrates: the main difficulty is in articulating the nature of the difference. The parallels in bodily physiology, in the nervous pathways transmitting signals from damaged tissues, and in the behaviour in response to such signals, make it improbable that man is grossly more sensitive to 'pain' than other warm-blooded vertebrates. In fact, in acuity of some sensory perceptions he may be inferior: the dog's sense of smell, for instance, is far more acute than man's. But there is one general area where man is outstanding, not often pointed out in this context: that is *his capacity to accumulate his experience and his solutions to the problems he has encountered, by means of the spoken and particularly the written and printed word.*

One can recognize that other beings can, in a sense, accumulate their past: a coral reef or an anthill are examples, but these are mere aggregations of the products of the same activity. The process of evolution might, too, be construed as a process of accumulating solutions to environmental pressures. The process however is slow, spreading over the millennia. One could argue, too, that the rough generalization that 'ontogeny repeats phylogeny' shows the accumulated record of our evolutionary past, for instance in the vestiges in our bodies corresponding to gills or a tail. Yet these vestiges are functionless, and do not give us the capacity to breathe under water or to swing with limbs free from a tree, in the way that man's past discoveries can continue to be put to his service. Much evolutionary change in fact represents the exchange of a less advantageous capacity for a better one, and much of the past history is deleted.

But at best, all that we can discern are the merest vestiges of the beginnings of those qualities which led to man's accumulative capacity; and they are trivial compared to the consequences of that capacity. At the heart of it is not merely accumulation but communication. This is particularly striking in medicine. The possibility of human anaesthesia by ether was discovered in 1846 in Boston, and within a few weeks it was being practised across the Atlantic. At once new anaesthetics, improvements in technique, new applications began to flow; and the fact that surgery could now be painless opened the way in its turn to antiseptic and aseptic techniques. Röntgen's X-rays spread with a similar rapidity. The manner in which each of us builds on the results of other men's work is peculiarly vivid to the scientist, and is crystallized in the bibliography appended to his papers, the eagerness with which he reads his journals, the proliferation of abstracting journals (and even abstracts of abstracts). More generally the proliferation of text-books, instructional handbooks, indeed the whole educational system, expresses the same establishment of past achievement as a stepping-stone for new achievement to come.

The effects of accumulated knowledge are so familiar, so much taken for granted, that they are difficult to articulate. They are evident in the diversity and extent of man's artefacts, his technological power, but above all in the records of his activity—in the libraries of the world (and soon in the computer stores). It is because of that written record that the schoolboy can master in a few weeks the calculus that Archimedes began with his 'method of exhaustion', and whose formulation taxed the greatest brains of the seventeenth century. The musician can see the development of Pythagorean harmonies, read Rameau's treatise, explore the works of Bach, and have a tutorial with Hindemith, before he strikes out on his own. The naturalist does not start, like Adam, naming species from the beginning, but draws on the descriptive work of thousands, as well as on the classificatory guidance of Linnæus and his successors. The engineer need not invent steel or concrete, or the steam, diesel, or jet engines afresh, but can envisage yet new materials or new ways of harnessing energy. The outcome, of course, is that man has an incomparably greater knowledge of the world than any other species, and to match this an incomparably greater power, and prospects of yet more. Whatever vestiges there may be in animals of the qualities that have allowed all this, the fact is that they are fulfilled only in man.

For the moment, I intend to argue only that man, as we see him and the results of his activity, is so radically different from animals that he is to be regarded differently. We need not regard his activity as 'progress'; we need not yet consider questions of value, nor the significance of the fact that he alone feels 'responsibility' for the world around him. Our only conclusion now, in common with general thinking, is to place an operational upper bound between the human species and the rest of the animal world.

The lower bound: vertebrate–invertebrate

Where do we place the lower bound? This has always presented difficulty. From the start, it was the animals that man had domesticated or made his companions, especially the dog, cat, and horse, which aroused especial concern; and from this arises general agreement that all warm-blooded mammalian vertebrates must be included. Such was the feeling in 1875–6 about the chosen favourites, that the initial proposals were to ban absolutely any experiment on cats, dogs, or equidae (horses, donkeys, mules);[7] but in the final legislation, experiment on these was allowed under special certificates. At that time experiments on rats and mice were fairly limited. But soon they became the commonest subject for experiment, and it was recognized that the arguments that they can suffer are no less strong than they are for cats and dogs: this has led to more attention being paid to them. Argument continues, however, about how far domestication, or biological evidence of the extent of brain development, should influence the level of concern. Yet more difficult has been the category of cold-blooded vertebrate animals, particularly frogs and fish. It might indeed be the case that no form of consciousness is possible at these lower temperatures: brain development certainly falls far short of that in the mammal. Yet there remain sufficient similarities, for instance in the details of brain structure, for the contrary to be argued. The original draft of the 1876 Act excluded 'cold-blooded' animals from its provision, and it was only in the committee stage in the House of Commons that a new distinction—that between 'invertebrate' and 'vertebrate'—was adopted.[8] This means that the frog, but also the tadpole, newt, and even the smallest fish, are all included under the Act: it is therefore a reasonably cautious dividing line. It was also one that the Littlewood Committee reconsidered in 1965, and saw no reason to change.[9] Some would think inclusion of tadpole or newt unnecessary, others that the range should be widened to include some invertebrates such as the octopus or lobster.

The centre of the problem appears to be the condition required for consciousness, particularly consciousness of pain. One might envisage animals existing into whose consciousness pain never enters. But this seems improbable since the physiology of pain as seen in ourselves and other mammals[10] is evidently a mechanism for warning of bodily damage, even if the signals are sometimes inappropriate. Painful stimuli override all others: the electrical signals they arouse appear very widely in the brain and not, like touch or vision, in select areas; accommodation does not occur as it does, say, to touch (illustrated by the way you cease to feel the spectacles on your nose after a short time); if anything, the sense of pain 'recruits'—that is, gets worse with time; there is a particular urgency associated with the sensation; the reflex physical response to a painful stimulus is very stable, and one of the last to be totally abolished in anaesthesia, so that the 'corneal reflex' (a blink on touching the cornea) is a useful test of deep anaesthesia; and the conditions in the tissues that give rise to signals in the nerves concerned, where they have been analysed, are in fact those where damage to cells is impending but still reversible. Given that pain is biologically so important, we would expect an animal that had any consciousness at all to be conscious of painful stimuli. Although the biological function of consciousness is very obscure, it is hard to believe that it is not in some way related to controlling actions by the organism. If that is accepted, consciousness as a mechanism of control which yet was unresponsive to signals of bodily damage could hardly be expected to be retained in evolution. The question therefore can probably be reduced to that of the conditions required for the presence of consciousness as such.

Three main ideas seem to underlie people's thoughts as guides to the probability of some state of consciousness and thus potentiality for suffering: level of complexity of behaviour; analogies to man in nervous-system structure; and size of nervous system. Some examples may serve to show how inconclusive such ideas can be. First, the highly complex behaviour of the ant or the bee is associated with a tiny nervous system, which few would believe capable of consciousness; it is clear that complex behaviour, whose neural substrate we certainly do not yet fully understand, is not sufficient evidence. Secondly, consider a man who has suffered a transection of his spinal cord. He has now lost all consciousness of the parts of the body supplied by the cord below the transection. But after 'spinal shock' has passed off, that severed piece of spinal cord is still capable of mediating a considerable range of responses to stimuli, and under the microscope it would be found to consist of a considerable mass of nervous tissue, displaying very great complexity of connections between the nerve cells. Despite these reactions and this complexity no one, I think, would suggest that the transection has created another 'consciousness' centred in this separated piece of cord. Thirdly, consider an anaesthetized man. Consciousness is lost and suffering cannot occur; yet a wide range of reflexes still operate, response by movement still occurs to painful stimulation, brain waves can be recorded, and with light anaesthesia seemingly purposive movement can be displayed. Analogous phenomena are seen, of course, in a bird or frog deprived of its brain. Such examples show that neither responsiveness to stimuli nor complex behaviour nor simple mass or complexity of neuronal tissue are enough to determine whether consciousness is present or not.

But if we try to look for some unequivocal basis for consciousness which we could use as a test for its presence in an animal, we will look in vain. No such test exists—yet. And while it is a tame conclusion, we can conclude little more than that the answer lies in some particular pattern of organization. It may indeed be that for that pattern to be possible, a minimum number of nerve cells, or sets of nerve cells arranged in co-operating groups is required, just as a computer cannot perform particular functions without memory, programmes, and data banks of a sufficient size. Thus there could well be a minimum brain size for consciousness to appear. But, as with a computer, that size will be necessary but not sufficient, and the function will be possible only when its organization is appropriate. For the moment, then, it seems that we have to proceed in almost the crudest way possible, aggregating our knowledge of brain structure, behaviour, and relationship to other animals as best we can.

In trying to decide what to include in the term 'animal', therefore, we must conclude that, for our present purpose, the upper bound falls short of man, and the lower bound is placed at the division between vertebrates and invertebrates. It must be restated that this does not rest on any rigorous logical division, but on operational 'cuts' in an apparently continuous spectrum such as we are constantly making in other walks of life. The distinction is important: those who wish to argue that the upper bound is irrational because no one can point to an absolute dividing line between man and the other primates, must go on to argue that the lower bound is equally irrational and equally has to be abolished. On that approach, if animals were to be treated as though they were man, then insects, and perhaps even plants, bacteria, or clay, would have to be treated as though they were animal.

Discrimination between animals

The defining of the meaning of animal leaves the question of differentiation between them, mentioned briefly above. One attractive source of such discrimination occurs when a particularly careful study is made of some individual species. This invariably brings to light hitherto unsuspected capabilities, seeming to raise that species above its neighbours. Yet we should recognize what could be called a 'fallacy of exaggerated attention'; it would not be accurate to set that species above its neighbours unless the same devoted attention were exercised throughout. The difficulties of discrimination are obvious, as between, say, a rat and a rabbit. Yet few would have difficulty between a chimpanzee and a tadpole. Many feel, too, that the relationship set up with domestic animals should be recognized; others, that even if we are fonder of cats and dogs, the rat and mouse can suffer equally. Bentham's question 'can they suffer?' may still be asked. There is no simple answer, but a general approach could be to think in a general way of 'lower' and 'higher' forms of animal, just as (to use the previous analogy) we think of 'lighter' and 'darker'.

Because of the precision required in legal drafting, it is hardly possible to include statutory provision for the rough-and-ready idea that experiments should where possible be made on 'lower' rather than 'higher' forms. But Article 7 of the proposed European Convention, which it is hoped will produce a general raising of standards internationally, offers scope for this approach:[11]

When a procedure has to be performed, the choice of species shall be carefully considered, and where required, be explained to the responsible authority; in a choice between procedures, those should be selected which use the minimum number of animals, cause the least pain, suffering, distress, or lasting harm and which are most likely to provide satisfactory results.

This is a wisely framed prescription. It carries us back from semantic problems to the essential objective: to try to choose that animal with which the suffering will be least in the obtaining of satisfactory results.

There is one last point which may provide some reassurance to those uneasy about incorrect choices. In Britain, the welfare of animals is also covered by a variety of other legislation, particularly the Protection of Animals Act of 1911, which brought together a considerable body of previous legislation.[12] Its range is considerably wider so far as species go, but it differs from the 1876 Act in that it depends on *actual suffering having occurred*. It is not always understood that the 1876 Act is fundamentally preventive, prescribing many conditions so as to avoid or minimize suffering; and conviction for an offence could occur without any actual animal suffering taking place.

One hopes that this aspect of legislation about animal experiment will always be maintained. But the existence of the more general 1911 Act is valuable as a further protection, particularly relevant to any inadequacy in the placing of the lower bound in legislation dealing with animal experiment. The 1911 Act is there to deal with unjustified suffering in animals generally.

Summary

1. If we review the 'scale of creation', it presents itself as a continuum in which we can discern no *absolute* dividing line separating creation into different categories, whether we use movement, growth, reproduction, self-protection, response to stimuli, sensitivity, learning, apparent purposiveness or evidence of pain-sensing mechanism, or evidence of suffering, problem-solving, consciousness, or fear.

2. Yet we readily acknowledge that our concern for the well-being of a cat or dog is different from that for an ant, grass, or clay, and that the problem is a familiar human task of making operational 'cuts' in a continuous spectrum.

3. An 'upper bound' can be made between man and other animals, because of man's capacity to accumulate experience, and to build on it, by means of the spoken and particularly of the written and printed word. It is this that has given him a greater knowledge of, and power in, the world than any other species, and prospects of yet more. The vestiges of the qualities that allow this may exist in animals, but are fulfilled only in man.

4. A 'lower bound' is commonly assigned either at the distinction between warm-blooded and cold-blooded creatures, or between

vertebrates and invertebrates. It is argued that any being that can move and interact with its environment will have some system warning it of bodily damage, and that this warning will be present in any consciousness it experiences; for it is impossible to believe that a creature which possessed a consciousness which was *not* responsive to adverse circumstances would survive in evolution. The essential question, in trying to place a reasonable lower bound, may then be reduced to estimating the extent (if any) of such consciousness.

5. If this operational approach is *not* adopted, there appears to be no reason for distinguishing between the care due to minerals, bacteria, vegetables, animals, or man: all should be treated alike.

3

WHAT IS AN 'EXPERIMENT'?

Observation, intervention, and analysis

It is vital to realize that we are discussing a process of discovering something that, before the experiment, was not known. It is this that provides the great objection to any legal provision which requires the beneficial outcome of an experiment to be precisely specified. We need to be clear about the nature of experimental work. It is not the same as new 'discovery', in the widest sense of the word, for that goes beyond the scientific experiments we are here concerned with. For instance, although if you have solved an anagram, or found the roots of some awkward equation, you have discovered something you did not previously know, yet it did not involve experimental work. Nor is the world of experiment simply synonymous with that of science; much scientific work is purely observational, and it is sometimes urged that the animal experimenter (or medical research worker) would do better to *observe* nature more and *intervene* less. To illustrate these points it is worth considering a simple example of scientific investigation.

(1) We wish to know how two new barbiturates compare in the duration of the sleep they produce. Two groups of six male mice are selected, weighing 20–21 grams, and each injected with a suitable dose intraperitoneally (into the abdominal cavity). They are placed in a warm environment, and the time is measured at which the ability returns to right themselves within 10 seconds if turned on their backs. The difference in the average sleeping times of the two groups gives the desired comparison, and a statistical test (using the standard deviation of each average) allows us to assess the reliability of the test.

(2) Following further study of their satisfactoriness, these barbiturates are to be compared in hospital patients having difficulty in sleeping. Patients weighing 50–51 kilograms cannot be ordered from a supplier, so we depend upon the chance of suitable patients being admitted to hospital, seeking to obtain two groups, made up of similar age, weight, and sex distribution, to each of whom a suitable dose is given. Instead of testing their

righting reflex, nurses record their sleeping time and the patients report their own estimate (often very different).

(3) It is known, however, that mice normally sleep by day, and their feeding and social activity takes place at night; so the first experiment can be regarded as incomplete and a picture of the diurnal activity of these particular mice is needed. A number are set aside, therefore, of the same type as before, in a cage with adequate food and water; and an 'activity meter' is arranged (e.g. a radar field which records every movement, adjusted so as to reject respiratory movement but to record anything more extensive), and left running day and night for a few days, with a continuous recorder. This gives a graph of the pattern of activity for this group. Suitable times of day can now be chosen to repeat the earlier tests.

(4) Stimulated by this, one might wonder about the pattern of sleep in hospital patients. They are unlikely to be willing to be placed in a group of six in an enclosed space, with food and pellets, for a few days. So the nurses could simply be asked to record their sleeping behaviour without any comment or intervention. A series of very varied patterns results, which may be described by an average, or by ranges, or simply presented as a collection.

These four examples help us to escape from semantics, and to tease out some essential features. In each case there was a 'question' to be answered. All the procedures were 'observational', so that this word alone is not very instructive. They all produced something that was not known before. But the four runs differ in three important respects. First, in (1) and (3) we deliberately intervened by setting up an experimental procedure and its raw material, selecting subjects (mice) by particular criteria; in (2) and (4) nature was allowed (by the chance of admissions to hospital of a series of individuals) to provide the raw material, and our role then was not that of intervention but only of selection. Secondly, in (1) and (2) we intervened by giving a treatment, whereas in (3) and (4) we only observed. Thirdly, in (1) and (3) a *comparison* was made between two types of treatment; in (2) and (4) we ended simply with a pattern, a description.

These runs, elementary as they are, fail to do justice to the richness and diversity of scientific work, but they roughly straddle the extremes of scientific study, from simple observation of 'nature's experiments' at one extreme to various forms of intervention at the other: providing the test objects, formulating an experimental procedure, and introducing some treatment or other 'perturbation' (to use a physical term).

It was the introduction of deliberate intervention into the study of natural phenomena, as opposed to observation, which so inspired the investigators of the seventeenth century onwards, and made them talk of 'the way of experiment'. Galileo, Torricelli, Gilbert, Boyle, Hooke, and their peers showed what a wealth of new knowledge could be gained if, instead of just observing natural events, you set up your own test to make events happen under your own control, choosing materials, procedures, and comparisons. That a springy material bent further if more heavily loaded must have been noticed by most people over the centuries; but it needed Hooke to compare the magnitude of the movement with the magnitude of the load ('Ut tensio sic vis') to reveal the simple law. Similarly, once methods of producing a relative vacuum had been worked out, Boyle might simply have enumerated all the effects on objects placed within it; and indeed to some extent that is what he did. But he also compared, for instance, how volume increased as pressure fell (as others did conversely by testing how much a volume decreased when subjected to greater depths of water), revealing a very simple relationship, Boyle's law, that pressure multiplied by volume is a constant. The difference between random observation and deliberate questioning is illustrated by his placing of a bell, struck by a mechanical device, within the evacuated chamber of his 'engine'. Why should he do this? It was to test whether air was responsible for carrying sound. The gradual fading of the sound, and its return when the vacuum is released, is one of the most simple and satisfying of his experiments.

This concept of an intervening deliberate simplification is of importance today, when biological and medical science is accused of being too analytic, of not thinking 'holistically', of ignoring the rich complexity of particular individual instances in natural life. If one turns back to Hooke and Boyle, can anyone believe that either could have arrived at their laws by mere inspection of the casual bendings or movements of the world around them? Nor does the point stop there; in fact, both laws are inaccurate and incomplete. The tension of a spring also depends on temperature and on the *rate* at which it is stretched; and the change in volume of a gas, caused by change in pressure, in Boyle's experiments would also have been sensitive to temperature, to the particular gas used (at the extremes of pressure), to gas dissolved in the walls of his containers or lost by diffusion, to water vapour forming or condensing, and many such factors. But until the first primary law was discovered, not even a start could be made in disentangling these other factors (leading in turn to the wider Van der Waals's law, and to new knowledge of intermolecular forces, to Henry's law of gas solution, to Fick's law of diffusion, and to the discovery of water-vapour pressure, dew point, and hygrometry). After centuries of observation, the discovery of how much could be done by simplification in analytic experiment constituted an intellectual thrill whose impact can still be felt in the writings of the time. Biological and medical events are far more complex than these relatively simple phenomena; and the more complex they are, the less hope there is that mere inspection can reveal any pattern, and the greater the need for analysis. Unless one abandons hope of *any* pattern, *any* generalization, *any* principle, and leaves each event or case individually isolated, bearing no relationship to any other, the complexity of such phenomena merely reinforces the need for experimental analysis, for disentangling the various processes and their interaction.

The unknown and the unexpected

There is an extra dimension to the possible outcomes of an experiment. What happens quite commonly is not an answer to the question put, but something quite unexpected. An example would be to find in the experiment outlined above that one of the new barbiturates did not produce sleep but excitement or convulsions (as is the case for one or two of them). A remarkably similar instance is that of a simple chemical derivative of ether, the classical surgical anaesthetic, namely hexafluoroethyl ether, which proved to be so reliable a convulsant that it was used clinically as an alternative to electroconvulsion therapy. We should not be surprised how often this occurs, for it would only be if we had a full prior understanding of all the mechanisms in play in a living organism that we could be sure that we had put the right questions. It is, in fact, the mark of a shrewd investigator that he is on the look-out for the unexpected, and gives it a chance to appear and be recognized. As Pasteur remarked, 'Chance favours the prepared mind', and many others, whether humble or elevated in reputation, would agree.

Another example is worth recalling. As a young man, H. H. Dale (later Sir Henry) went into the pharmaceutical industry, and was asked by his employer, Henry Wellcome, to look into the pharmacology of a traditional remedy for inadequate uterine contraction in or after childbirth, namely ergot, a fungal growth on rye.[1] Part of this work involved checking samples for their activity. Such natural material is liable to bacterial contamination, and as a result may contain a variety of products of bacterial metabolism (as does also, for instance, cheese). Although these contaminants were not related to the active principle of ergot itself, which was being tested on the blood pressure of anaesthetized animals, their effects—varying from sample to sample—aroused Dale's curiosity. The dividend was remarkable: first, some substances resembling adrenaline were found, which gave rise to great advances in our understanding of the sympathetic nervous system; secondly, a substance called histamine was identified, which opened up other fields of work, namely in circulatory shock and what came to be called allergy. Thirdly, a substance called acetylcholine was discovered. This attracted Dale's special attention because he found that, after giving a dose that had apparently produced a fatal fall in blood pressure in an anaesthetized animal, the blood pressure nevertheless recovered fully in a few minutes. He reasoned that to be disposed of so quickly there must be something in the body equipped to do this, so that acetylcholine might have a deeper significance, as indeed was the case: it proved to be the first neurotransmitter whose role and identity was unequivocally established, and is now known to mediate the nervous control of our muscles, secretory glands, and important pathways in the brain. Thus a rather *ad hoc* practical question about ergot opened up three new areas of physiology and pharmacology. That brings us at once to the question of the distinction between applied and fundamental research.

'Fundamental' and 'applied' research

Within experimental work, the distinction between 'fundamental' (or pure or basic or academic) and 'applied' (or practical or useful or mission-oriented) research can become important. For it may carry the implication that fundamental research is regarded as merely 'wanton curiosity', while it is only 'applied' research with practical benefits in view that could readily justify animal experiment. The distinction has become more common since its articulation in the Rothschild Report of 1971.[2] There, 'fundamental' work was regarded as simply adding to knowledge, whereas 'applied' was characterized as suitable for a detailed contract between investigator and customer. It was suggested that government support for fundamental work might reasonably amount to 5–10 per cent of the funds available, the remainder being suitable for contractual handling. The 'customer–contractor principle' may owe some of its attractiveness to its apparent definitiveness. It seems easier to judge whether a contract would ever be agreed for the undertaking of some investigation than to decide whether it is 'applied': being a contract, the 'use' of the work would be defined in some sense. It was therefore a proposal that appealed to administrators and to those seeking to make science meet their standards of 'accountability'. But it also ran across the grain of a great deal of scientific work, especially a great deal of medical research, which is of a 'fundamental' type yet deliberately entered into to help create a baseline or framework from which applied work could grow. Such work the simultaneous Dainton report had, more perceptively, identified and named 'strategic'.[3]

It is, however, still worth making a distinction between research seeking addition to knowledge, and research seeking practical benefit, simply to clarify discussion. In practice, the distinction becomes hopelessly blurred, since practical research commonly produces major additions to knowledge, and fundamental research commonly yields important practical benefits. Nor do investigators restrict their objectives to one or the other, but can often change in mid-stream, as Dale did, from a rather practical study on ergot to what was (then) an academic study of the acetyl ester of choline. Strategic research naturally aims at both. There is the fact, too, that some (including myself) have not been able to identify *any* contribution to knowledge from which some practical benefit could not be envisaged.

If one is concerned over the justification of the use of animals for some set of experiments, it is possible sometimes to say that their purpose is primarily to increase our understanding and to remove ignorance about some bodily process, without really being able to say how it might be useful in other than the most general terms. At other times there is a more distinct practical goal, sometimes very specific as in testing toxicity or in choosing between possible new drugs, or in following the metabolic fate of a substance in the body. In general we have a continuous spectrum with each component in varying proportions. But the distinction has some value in that the warranty for the use of animals is probably seen as different in the two cases. Later, therefore, we shall distinguish between benefit by addition to knowledge on the one hand and practical benefit on the other.

Summary

1. Experimental work is concerned with *what is not yet known*. This immediately implies that the outcome of such work cannot be specified in advance.

2. New knowledge was won over the centuries by reflection and by observation. But the great discovery, termed the 'way of experiment', around the seventeenth century, was of what could be gained by deliberate *intervention*, choosing test material and test procedures so as to simplify and disentangle various processes. It is the adding of deliberate analytic experiment to the observation of 'nature's experiments' that characterizes the experimental approach.

3. The outcome of an experiment is commonly not merely unknown but also *unexpected*.

4. The distinction between 'fundamental' and 'applied' research cuts across the grain of actual scientific practice: the term 'strategic' research more accurately describes that large body of medical research work which deliberately seeks to do work of a 'fundamental' type that will provide a framework within which 'applied' work may grow. But although the distinction is blurred in practice, it is worth keeping in mind to clarify the discussion of the benefits of experimental work, as flowing from new knowledge or new use.

4

THE ETHICAL QUESTIONS

Is it right in principle to do to an animal what you would not do to a man?
Is it right to do to a man what you could do to an animal?
Is it right to do to a plant what you would not do to an animal?
Is it right to do experiments on higher animals which could be done on gravely handicapped humans?
Does not an animal have a 'right' to its natural life?
How can an animal have a 'right' if it has no responsibility?

The purpose of this chapter is primarily to consider certain arguments of principle. If strongly pressed they can lead to the absolute antivivisectionist position—that experiments on animals should be absolutely prohibited. If less strongly pressed, the issue ceases to be absolute, and the question of some sort of balance arises. This involves detailed scrutiny of the gains and losses to animals and man resulting from animal experiment and some decision as to how that scrutiny should be conducted.

The reader is asked to review the questions at the head of the chapter for a moment. They are a small sample of those that can be asked: they show, incidentally, that the way a question is put can place the onus of proof in one or other direction. The answers depend, in one way or another, on one's knowledge of, and view about, the relationship between man, animals, and the rest of creation. The *general* issue now is whether or not there are differences between men, animals, and plants which justify different treatment. Animal experiments may involve the costs of pain, suffering, loss of life, or interference with 'normal' life, so we must also establish what the cost of the treatment is.

Pain and suffering

It is obvious, as one reads the older antivivisection literature, that pain took pride of place. The views expressed were exceedingly trenchant and often absolute, calling for abolition, not mere control. As we search for the reasons behind it, we might expect an absolute opposition to any suffering whatever; but in fact we often find acceptance of many veterinary practices of the day and of a good deal of temporary animal suffering for what were regarded as good reasons. The real motive force is revealed as horror at the contemporary or past accounts of scientists' experimental practices.[1] It was not that *no* suffering was the ideal, but rather that *this* particular suffering was intolerable, even if it was to advance knowledge. The idea of a balance between suffering now and preventable suffering in the future is implicitly acknowledged, for the arguments always followed two lines: on the one hand, the expounding of experimental horrors; on the other, a depreciation or refutation of the knowledge or prospective benefits to be gained, which is only relevant if some balance is to be struck. Many of the cases cited are indeed horrifying to a modern mind. But judging the past by our present standards is a misleading activity, and we need, before making a judgement as to what was or was not justified, to review the suffering at that time in homes or hospitals, in asylums, on the battlefield, in the factories, and in the countryside. When human life and suffering was cheap, animal life would hardly be valued higher.

Echoes of past examples of experimental work still appear, although present circumstances make them almost irrelevant, apart from their historical role in shaping institutions and traditions. The modern reader should be on guard lest information from that past is being paraded as current practice, just as he needs to be wary of work in other countries being presented as typical of this country. Some of the past literature, particularly the anti-vaccination propaganda, makes sad reading to a generation that has at last seen the abolition of smallpox.

One may guess that the introduction of anaesthetics and improvements in experimental practice led to the subsequent change in attitudes, since many experiments now clearly involved very little or no pain. A more absolute argument developed, the only *absolute* one I have encountered; namely the simple contention that it is absolutely wrong to inflict suffering on any living creature. 'Suffering' is a more general term than pain, and could include anything from distress to death, and 'living creature' may include a narrow or wide range of animals. An abolitionist position may be held while fully recognizing the cruelty of the natural world, its predators and victims, natural disasters, diseases, and the ordinary accidents of life; for it can be argued that, at the very least, one should keep one's hands clean by not adding to it in any way, so far as one can avoid it. This approach is accompanied, of course, by vegetarianism and other logical consequences. It is clearly a tenable attitude, if the vegetable kingdom is not included in that part of the living world whose existence or well-being is given absolute value. It may also be defended on economic, ecological, or health grounds, and may accompany a belief in herbal remedies. The price to be paid, of course, by this avoidance of particular sins of *commission* is (to other eyes) the growth of sins of *omission* and the rejection of all the benefits to animal and human life that have flowed, and can flow, from knowledge gained about the animal body, which will be described in Chapter 5. It would allow people or animals to die of preventable disease; just as according to certain religious beliefs it is right to allow a mother to die of haemorrhage after childbirth rather than give a blood transfusion. Other beliefs may support this position, such as the belief that it is good to abstain from action and to escape from the world of striving. That would entail a stepping aside from the living world as we see it, in which, at all levels, activity, effort, struggle, and seeking the limits of endurance are an inherent part. Total abolition also cuts off a large field of human biological enquiry, and accepts a corresponding permanent ignorance.

We may ask *why* should it be felt absolutely wrong for a man to kill or inflict suffering of any sort on any living creature? It seems likely that it is simply one of man's responses to the suffering around him because of his capacity to project his emotions to other beings, to 'sympathize'. If we think of the conditions of human life at the time of the birth of Buddhism, we can respect this impulse. But one point arises. How does it come about that a man can be *blamed* for taking animal life, when we do not *blame* a cat for killing a mouse, or a dog for worrying a sheep (however much we blame the owner)? We meet again a difference between men and animals, that it is only humans that are governed by words like 'right', 'wrong', 'ought', 'blame', 'responsibility'—words which express moral concepts. How it comes about that humans have a moral sense is beyond the scope of this book. But every time such terms are used, the difference between man and animals is reaffirmed.

Not many people take such an absolute position. But suffering remains an essential factor in the argument, expressed in a commonly quoted passage from Jeremy Bentham, written in 1780:

> The day *may* come when the rest of the animal creation may acquire those rights which never could have been withholden from them but by the hand of tyranny. The French have already discovered that the blackness of the skin is no reason why a human being should be abandoned without redress

to the caprice of a tormentor. It may one day come to be rcognized that the number of the legs, the villosity of the skin, or the termination of the *os sacrum* are reasons equally insufficient for abandoning a sensitive being to the same fate. What else is it that should trace the insuperable line? Is it the faculty of reason, or perhaps the faculty of discourse? But a full-grown horse or dog is beyond comparison a more rational, as well as a more conversable animal, than an infant of a day or a week or even a month, old. But suppose they were otherwise, what would it avail? The question is not, Can they *reason*? nor Can they *talk*? but, *Can they suffer*?[2]

It is an interesting passage: as well as focusing on suffering as the essential criterion to be used, it introduces the comparison of animals with infant humans, the idea of animal rights, and the idea of an animal liberation comparable to that of release of the negro from slavery. With any such quotation we need to distinguish between the argument in itself and the authority conferred on it by the name of the author. It is, therefore also worth citing what Bentham says a little earlier:

there is very good reason why we should be suffered to eat such of them as we like to eat; we are the better for it, and they are never the worse. They have none of those long-protracted anticipations of future misery which we have. The death they suffer in our hands commonly is, and always may be, a speedier, and by that means a less painful one, than that which would await them in the inevitable course of nature.[2]

So, too, one finds that while one may gain inspiration from J. S. Mill's plea for personal liberty, yet one also finds him presiding at the India Office at the peak of the opium trade with China, and a vehement opponent of State education. The quoting of great names is a double-edged pursuit: unless one accepts the whole bundle of their opinions, it is the cogency of the argument itself, not the fact that a particular great man advanced it, that is relevant.

Sentiency, purposiveness, and self-enrichment

So, let us set aside authority as such, and focus on the specific suggestion that Singer,[3] following Bentham, has urged most strongly. This is that the essential criterion is 'sentiency' (a convenient shorthand term for 'the capacity to suffer and/or experience enjoyment'). It is this that gives rise to 'interests' of which we must take account. Whenever 'sentiency' is present, there is an 'interest' which must be weighed equally with other interests, such as human welfare, knowledge, and the like. The argument is not for equal treatment but for equal consideration. It leads to a classical utilitarianism, in which the consequences of any action are to be calculated in terms of the good to be derived against the suffering produced. In a general sense, this sort of approach would be used by many, and the real difficulties or conflict arise in the weight to be ascribed to the 'benefit', the measurement of the 'suffering', and the balancing of the resulting incommensurables.

But this particular calculus leads to more controversial ground. Sentiency cannot be denied to animals, so we must face the suggestion that a higher animal is at least as 'sentient' as, and possibly more so than, certain human beings—such as a baby, or the hopelessly senile, or an individual with a very severe handicap, or people with certain forms of mental disorder, or a gravely ill patient. It would also be true that any subject, human or animal, who by treatment with pain-killers, or by suffering from a rare congenital freedom from pain (dysautonomia), would also have to be regarded as correspondingly insentient. The last point has prompted an alternative proposal: that it is because 'animals and humans are so organized as to be *purposive* creatures with various desires, drives, intentions and aspirations', that animals acquire 'rights'—to be left alone, to pursue ends, to be allowed to live.[4] The extent of interference with 'purposiveness' becomes the key. A further variant arises in a slightly different way, from a search for some defining quality most likely to do justice to any fundamental difference between human and animal life; this is 'capacity for self-enrichment', as exemplified by things which make human life valuable:

> the pleasures of friendship, eating and drinking, listening to music, participating in sports, obtaining satisfaction through our job, reading, enjoying a beautiful summer's day, getting married and sharing experiences with someone, sex, watching and helping our children grow up, solving quite difficult practical and intellectual problems in pursuit of some goal we highly prize, and so on.[5]

Consent

A rather different argument is that it is wrong to do experiments on animals, because they cannot 'give their consent'. The discussion in Chapter 7 on the assessment of pain and suffering outlines the various ways in which evidence about these could be obtained in animals, and it is clear that they could indicate suffering. But while it is possible to know that a man is willing or unwilling deliberately to override his expression of reaction to pain or suffering for some good cause, we cannot know this about an animal. From this comes the sense that it is 'unfair' to inflict deliberate suffering on an animal. That very statement, however, points to some radical difference between men and animals: both the conception of 'fairness' and the acceptance of it as carrying some moral obligation, appear to be specifically human. If this is the case, the implication appears to be, not that it makes animal experiment impermissible, but that it reminds man again of the nature of his moral responsibility in respect of animals. This principle has in fact long been recognized. The revulsion at the use of curare (which paralyses voluntary movement) as an 'anaesthetic', that led to a ban on its use in this way in the 1876 legislation, was precisely because it deprived the animal of the means of expressing its experience without preventing it from feeling pain. The conception of 'consent' in an animal offers considerable difficulty. We do not know if animals 'consent' to be domesticated, or to be killed for food or because they are strays, or for other reasons. Perhaps the underlying assumption of all legislation is that if any pain or suffering is involved then lack of consent is assumed: but it is hard to see much difference between saying 'We must minimize pain and suffering in animals because we assume they do not consent to it' and the simpler statement 'We must minimize pain and suffering in animals.'

The 'higher' animal vs. the 'lower' human

We have, then, 'sentience', 'purposiveness', and 'capacity for self-enrichment' proposed as criteria for counting cost. These need not be regarded as alternatives. To be as fair as possible, one could ask in turn about the cost in pain, in more general suffering, in frustration of purposiveness, and in loss of the capacity for self-enrichment.

One may note that the three criteria progressively introduce an element involving others. The first criterion, pain, is intensely individual. But failure to fulfil a 'purpose' or to achieve self-enrichment could well affect others in, say, one's family or in one's job. In counting the cost, therefore, we are becoming concerned with the total loss, not just the loss to the individual.

An important step in the argument then comes when it is claimed that some 'higher' animals would suffer a greater cost from experiment than some humans—such as infants, some diseased, some senile, and some congenitally handicapped—because of their superior capacities compared with such humans. The cost of experiment is therefore to be viewed as greater in the 'higher' animal than in these humans, and the special status of man is destroyed. This conclusion can be used in two ways: that because we devote such care to these human categories, we should raise our care of the 'superior' animals to at least that level; or that because we are willing to do certain things to these animals, such as killing to keep down their numbers and using them for food or for experiment, we should be willing to do the same to these particular humans. The second of these points of view raises in its most acute form the question whether the criteria that have been suggested are really adequate to do justice to what might be lost, in human or in animal life respectively, if humans or animals were subjected to scientific experiment.

Capacity for self-enrichment represented one author's best criterion (he was unable to accept criteria of a religious type)[5] for distinguishing human from animal life; and using it he had to conclude, as he and others had with sentiency or purposiveness, that there was no clear distinction to be seen; animals showed all three qualities. But 'capacity for self-enrichment', like the other criteria, is a pale caricature of the human quality, even if religious or similar ideas are set aside. In the first chapter it was argued that, although we should recognize a continuity in the scale of creation, nevertheless there is an operational 'upper bound' to the animal world between man and animal, based on his capacity to accumulate his experience by language, by the spoken, and (especially) by the written or printed word. It is this—humans interacting with other humans and building on other humans' achievements—that has opened up a difference so great, and still growing, as to have created a qualitative distinction in nature and value between the human and animal worlds. From this, too, comes man's capacity to consent or not to consent. If we are considering costs, we must also count the loss of the exercise of this accumulative capacity and its fruits.

At this point the reader may hesitate. We began with individual suffering: now we have extended beyond that individual to include others. Who is it that is 'counting' the cost, the sufferer or someone else? It seems that, in general, it must be someone else, namely the human (or humans) who are seeking to decide on the rightness of some action. A suffering animal or man, in isolation, is just that—isolated. It is only by sympathetic projection or communication that another being makes contact. That other being may 'speak for' the sufferer, but cannot 'be' them. How then is justice to be done to the individual? We could argue that one of the great pathways of ethical advance has been in this direction, in the sharpening up of 'projective sympathy' (the word 'projective' may seem redundant, but 'sympathy' can have a cosy self-wallowing flavour, and 'projective' stresses the mental movement outwards to other beings). So we ask not just 'What do I want or think?' but other less egocentric questions such as 'If everybody thought like this, what would be the result?' or 'If I place some other, arbitrarily chosen creature in my place so as to avoid all bias, using whatever anthropomorphic or human insights I can muster, what would be said?' It seems that it is from the aggregate of such essentially social thoughts that individual conclusions are reached. We may differ in our conclusions: thus Bentham's remark that if we eat animals 'we are the better for it and they are never the worse', is rejected by Singer. But it remains the case that, by the capacity for human communication, these individual conclusions are compared and integrated into social action.

We should now return to the comparison between the 'higher' animals and the 'lower' humans. Let us postulate some experiment: for instance, that a new and malignant strain of malaria has arisen, and that it is important to identify the nature and location of the intermediate forms it takes in the animal body with a view to appropriate drug development or other treatment. From clinical cases so far seen, infection of a subject is expected to cause a severe febrile illness, which may be fatal; but in any case the subject will have to be killed to examine the various tissues. We are to choose between doing such an experiment either on one of these lower humans or, say, on a chimpanzee. A number of comments can be made.

(a) Some of the cases cited of 'lower' humanity seriously underestimate their capacities. Thus a patient with spina bifida has been used as an example.[5] Yet spina bifida may be symptomless, or damage may be restricted to loss of bladder and anal control, or there may be some neurological deficit, mostly to the lower limbs. It is essentially a spinal disease, leaving the brain untouched, and compatible with a full range of human activity. If such cases are to be used, then severe cases of poliomyelitis are eligible too. Down's syndrome is regarded as a sufficiently serious human handicap for some to have the individual killed *in utero* by abortion. Yet those familiar with such individuals or comparable cases know that they can learn, teach others, earn money, arouse great affection, and join in family and communal life: their experience today, from their own efforts joined with that of normal humans, points the way to still further development of their capacities. We have to make similar remarks about the senile and the diseased. As to babies, while Bentham may not have found them 'conversable', the astonishing growth in organization of the nervous system and in mental capacity taking place in them from birth is as remarkable as anything going on in his own brain.

In the face of such comments, it may be said that the postulated case is more deficient than these, 'very very severely handicapped'.[5] But if no specific example is quoted, the serious empirical question arises: 'Do we know that there *are* any individuals so seriously handicapped as to be eligible, who have lost *all* personal human interaction, and who yet are alive? May this not be an 'empty set'? It is a point at which an empirical question appears to impinge on a question of philosophical principle.

(b) It is notable that the eligible human cases are all in some respect handicapped. In the past, such a case might have been a cretin (someone suffering from mental retardation caused by thyroid deficiency). Suppose it had been chosen for experiment in 1890. A year or two later, cretinism was preventable and curable if treated early enough, and a being with full human potential would have been used. Phenylketonuria, a biochemical disorder, is a more recent example. Indeed, the same consideration applies throughout. The baby grows up. The seriously diseased cancer patient today need not *totally* despair of cure. The possibility of transplant of nervous tissue to replace damaged brain areas is already achieving its first successes, and biochemical approaches to understanding senile dementia are very promising. The word 'hopeless' is sometimes used to describe such cases, yet just because it is *defective* human beings that are being postulated, scope for betterment is immediately created. We could indeed assert that strictly there are no 'hopelessly' defective human beings.

(c) Some would think it inequitable that humans already handicapped should be considered eligible for additional suffering or deficit. But as a practical point, it is unlikely to be regarded as scientifically satisfactory that whatever experiment is done should be done, not on a normal adult organism, but on the infantile, congenitally deficient, senile, or diseased.

(d) As a reviewer of *Animal Liberation* pointed out, there is a potential fallacy in the drawing up and ranking of different species by their 'likenesses'. He notes that Singer accepts evolutionary theory and uses it to support his case in saying that only a 'religious fanatic' can continue to maintain that *Homo sapiens* is separate and distinct from other species: but Hull then points out that Singer ignores important differences that exist between different units of evolution. He goes on,

> Biologists do not group organisms into species or order species on grounds of similarity. Two forms can be extremely similar and yet be classed as two separate species (sibling species). Conversely, a single species can be made up of extremely dissimilar organisms—for example polytypic species like dogs and dimorphic species like the birds of paradise. And species are grouped into higher taxa because of descent, not degrees of similarity. From the biological point of view, the relations that exist between races and between sexes of the same species are different in kind from those that exist between species. If the principles of evolutionary theory are to be taken seriously, there are excellent reasons for us to exhibit a greater moral commitment to a child than to a porpoise, even if that porpoise has a greater capacity for suffering or experiencing enjoyment. If this be speciesism, then make the most of it.[6]

The reader may well feel at this point that the obvious has been sufficiently laboured, and that he does not need convincing that it is better to use even a chimpanzee than a human if an experiment is of such importance that it must be done on one or the other. Singer evidently feels the same; but he, too, labours the point because his belief that there is no argument *in principle* for preferring experiments on animals to those on man underlies his doctrine of 'speciesism' with all its widespread implications. That doctrine, namely that to discriminate between human and animal species is as wrong as discriminating between races or sexes, rests on the likeness between man and animals.[3] The comments given above argue that the differences remain sufficient for clear qualitative distinction, even in the extreme cases. They serve, too, to answer the suggestion sometimes made that, if the benefits obtained by research on animals are really as substantial as is claimed, then that would justify human experiment such as that practised under the Nazi regime. The final genetic point made by Hull may indeed be sufficient for many. The difference between men and animals is considerably greater than that between the sexes or between races, and the analogy with sexism or racism is false. Man *has* travelled further than animals along an evolutionary pathway. It *is* possible to draw a clear and unequivocal biological distinction. But it has been worth labouring the point for another reason. One purpose of the animal liberation movement is to sharpen our sympathy for animals. It is hoped that this discussion will not come amiss in sharpening our sympathy also for the infant, the senile, the diseased, and the handicapped.

Animal 'rights'

None of this, of course, ought to weaken our projective sympathy for animals. But it leaves undiscussed the question of animal 'rights'. We can readily accept that sentience, purposiveness, capacity for self-enrichment, give a 'worth' to their possessor. In turn that 'worth' implies a duty of care among men, by virtue of that faculty by which we accept such a duty. If we accept that duty we acknowledge a 'moral worth'. It could also be argued that this gives a 'right' to the animal. But it does not necessarily follow that because something has a worth it therefore has rights: we may feel a worth in some beautiful object (such as a building, a picture, a piece of porcelain) and a 'duty' to look after it, but we would not admit that it had a 'right' against us. If we are looking for convenient terminology we could say (following Linzey) that we have a duty not 'to' the object but 'in respect of' it.[7] But once we do this, we break with the idea that duties necessarily imply 'rights'; so it would seem that the existence or not of rights has to be decided in some other way. Indeed, as we struggle with these ideas, which are the subject of a voluminous literature, the idea of 'rights' seems to become more and more confusing and perhaps less and less useful save for political purposes. Ritchie has argued that the idea of rights has originated in three ways.[8] To begin with, in an earlier more despotic world, 'rights' were those conferred on his subjects by whoever was in the position of ruler: they could almost be described as 'allowances' of what the subjects could have and do. Then the feeling that this was arbitrary and inequitable generated a claim for 'natural' rights, that the mere fact of existence created a series of claims, for instance to freedom, continued life, food, shelter, justice, education, and so on. But then the questions arise: What is their justification? What if claims conflict? How are just rights to be agreed? How are they to be met, and from whose resources? As Ritchie says:

> In the chaos of conflicting individual impulses, instincts, desires, and interests, we can find no stable criterion. We must go beyond them to the

essential nature of things. But what part of the nature of things is here relevant? Is it not simply—human society?

If we accept that 'rights' are a function of society, of the interaction between human beings, we may leave the argument there, recognizing of course that it is only the beginning of a much greater argument about social organization. We can indeed end up quite simply. Following Caplan we may distinguish 'moral agents' and 'moral objects'.[4] The former are capable of moral choice and can accept 'duties'; they can therefore make their claim to correlative 'rights'. It seems to be agreed that the only moral agent of whom we have knowledge is man, although we can reasonably trace vestiges of the origins of his moral capacity in animals.[9] The moral objects are those in respect of whom man has duties; but because they themselves have no duties, cannot make claims, and are not participant parts of society, they do not have rights. There are thus many more moral objects than moral agents. It is not always noticed that there needs to be some equivalence between rights claimed and duties accepted. If rights are claimed and there is no one on whom the duty of satisfying them can reasonably be laid, the claim is an empty one. But provided that both duties and rights are restricted to moral agents, i.e. to humans, the necessary 'balance of trade' between rights and duties is possible: the totality of claims justly made can be arranged to be covered by the totality of duties justly discharged. The duties of a moral agent, therefore, are to other moral agents as well as in respect of moral objects. The concept of animal 'rights', with its false analogies to racial rights and sexual rights, is replaced with no loss, and perhaps some gain, by the 'moral worth of animal life'.

The ethical background

His habit was to encamp near to the region of practice in all his philosophical enquiries.

W. E. Gladstone on Bishop Butler, author of *The Analogy of Religion*[10]

It may be felt that this discussion has not been an ethical one at all, but that it started with a particular ethical position, and worked out certain practical consequences. Certainly the language used has not been used with precision. But the great difficulty in such discussion is the lack of agreement that emerges for the layman as he reads moral philosophy. It is as though men do not agree, in respect of words like 'pleasure', 'good', 'right', 'fair', or 'ought', and their opposites, which of them derives from which, if indeed any of them have such an origin and do not have another source. On the other hand, if, to use Gladstone's phrase, one encamps 'near the region of practice', there is very substantial agreement; for instance, that it gives pleasure and is good, right, and fair, to diminish human and animal suffering so far as one can, and that one ought to do it. So the attempt has been made to centre on the region where the lines of thought seem, in practice, to come close to each other.[11] Various ethical systems appear in the role, not of prescriptive guides, but more of models, whose consideration can make the mind look at the issues from another standpoint, and sensitize it to interests that may have been disregarded. In the end, it seems that we have to balance matters up in our own minds, and then to balance them up again between each other. The despised strict utilitarian, whose calculus for achieving such a balance can lead to well-known absurdities, might feel somewhat contemptuous of this approach: for are we not then doing precisely the same as he, balancing moral principle, knowledge, practical benefit, and suffering, but with a method wrapped in mystery, even if dignified by the name of 'judgement'? That may well be the case; and it is a strong reason for refraining from dogmatism, and for trying to get the evidence and the arguments as clearly laid out as possible so that, at least, what is to be balanced constitutes some common ground.

Summary

1. Opposition to animal experiment was initially not on principle, but on the grounds that the animal suffering as perceived by the critics, was too great to justify the knowledge gained. With the introduction of anaesthetics and of other advances, the ground of debate has changed to that of questioning the justifiability of inflicting any suffering, or loss of life, on any animal, for any reason at all.

2. This critique rests on the argument that there is insufficient difference between man and other species to justify different treatment. Possible distinguishing criteria such as 'sentiency', 'purposiveness', and 'capacity for self-enrichment' have been considered, but found inadequate. It is argued, however, that it is the same characteristic of man that defined the 'upper bound', that also gives human life a special value: namely the capacity to

accumulate experience by spoken, written, and printed word, and all that flows from this, including the capacity to 'consent', to ask moral questions and feel moral obligations, and to feel responsible for animals.

3. A particular test case is the suggestion that it is as justifiable to do experiments on an infant, senile, defective, or diseased human as on a healthy 'higher animal'. This is rejected because (a) the capacity of these human cases is underestimated; (b) they have all been deliberately selected as humans defective in some way, so that there is the chance with all of them that they may cease to be so; (c) it is in general scientifically unsatisfactory to experiment not on a normal organism but on the infantile, defective, senile, or diseased; (d) the strict *biological* difference between man and other species is underestimated. The difference between man and animals is greater than between races and sexes.

4. The argument for animal 'rights' is reviewed. There seems no basis for a claim for natural rights for animals. But one can readily distinguish between humans as moral 'agents', and animals (amongst others) as moral 'objects' in respect of which humans may accept duties. The term 'animal rights' may be replaced with no loss, and perhaps some gain, by the term 'the moral worth of animal life'.

5

THE BENEFITS OF ANIMAL EXPERIMENT

In Chapter 3 we discussed the difference between 'fundamental' and 'applied' research, and concluded that in experimental practice research workers are constantly mixing these approaches or switching from one to the other. The labelling of a piece of work as one or the other is therefore often difficult. Nevertheless, in trying to assess the benefits won by animal experiment, it remains useful to distinguish the benefits of addition to knowledge as such, and the benefit of some practical gain. The distinction corresponds to Francis Bacon's '*experimenta lucifera*', or experiments shedding light, dispelling ignorance, and '*experimenta fructifera*' or experiments yielding fruit. Often it has been experiments shedding light, work that illuminated our general understanding or created a sense of pattern and coherence in the world, by such as Newton and Einstein, that have been most admired: more practical work (on steam-engines, mackintoshes, or papermaking) has been appreciated but relegated to the position of cart-horse compared to a Derby winner. In discussions about animal experiment, however, it has largely been the other way round. It is sometimes claimed that only practical benefits can justify the infliction of suffering on animals, and that the motive of satisfying man's curiosity is not enough. 'Curiosity', as a word, like 'inquisitiveness', has perhaps now acquired a rather gossipy, nosy, trivial flavour; but we lack a better one for that exploring human faculty that wants to diminish ignorance and increase understanding.

To assess these two types of benefit, then, we shall use a historical method, namely a 'test of deletion' whereby we first seek to identify those things which we now know and can do which were made possible by animal experiment. Then we suppose them removed, deleted from our knowledge for lack of the necessary experiment; then we see what the loss would have been. Some of the arguments against animal experiment are absolute, and they would have been equally valid at any time in past centuries, even thousands of years ago; so that the test of deletion may go back equally far. Other arguments against animal experiment are directed against particular aspects, such as psychological experiments, so that a more restricted deletion is necessary. Sometimes the argument is not absolute, but simply wishes greatly to reduce animal experiment at once; one has then to attempt an estimate of the quantitative reduction of benefit, rather than its total loss.

It seems that in the end it is necessary to use a historical method of this sort. As already stressed, it is an essential characteristic of scientific research that it is concerned with what is at present unknown, just as is the product of any other creative activity. Elgar, walking perhaps on the Malvern Hills, said 'Music is in the air all around, you just take as much as you want.'[1] It is the same with scientific discovery: it is all there waiting to be found, if only you can do it. But you cannot assert with the force of logical certainty what will be discovered, nor undertake to make a particular discovery. Consequently it cannot be asserted that in the future, if animal experiments were abolished or reduced, this or that particular thing would fail to come about. You can make some guesses but you cannot be sure. It is a difficulty that the animal experimenter has faced since at least 1840 when Marshall Hall was challenged about how he could guarantee that his experiments would indeed produce a 'good'. Like him, we can only look back, and show the sort of things that would have been lost from such a stop in the past. As in other matters, our confidence in future benefit is a function of the historical record.

Benefit from knowledge

We noted that sometimes new knowledge won by animal experiment is dismissed as 'merely' satisfying the curiosity of the experimenter. In a university, we take for granted the value of free enquiry and knowledge; and our task is to understand past knowledge, to interpret and preserve it, to teach it to the coming generation, and by building on it to seek to add to it. Although that very spirit of enquiry accepts that it is legitimate to question its own value, yet as with any question about long-accepted values it is not always easy to articulate the answer. We can start by emphasizing that it implies a removal of ignorance, and that the failure to win new knowledge means a perpetuation of ignorance (darkness, to continue Bacon's metaphor).

But let us apply our test of deletion, in a sort of 'think-experiment'. Suppose that no animal experiment had taken place for, say, 2,000 years, although scientific work on the inanimate world of course had continued unrestricted. Would the resulting ignorance and lack of understanding matter? People vary in the importance they attach to knowledge about ourselves and the world around us. But whether highly prized or not, we would now have a quite remarkably one-eyed view of the natural world. On the one hand, physics and chemistry, and their precursor and daughter technologies, could be fully advanced. There seems to be no discovery or insight in these subjects for which animal experiment was necessary. Galileo, Kepler, Newton, Dalton, Lavoisier, Faraday, the Curies, Rutherford, Einstein, or the engineers, the industrial chemists, the miners, the explorers, astronomers, geologists, meteorologists—all of these could still have gone forward. On the other hand, our knowledge of the functioning of ourselves and of the animal kingdom would be inferior to that of the ancient Greeks. To understand what that would mean, let us start with an experiment by the famous Greek physician Galen (AD 129–*c*.200), whom we can take as possibly the first significant animal experimenter.[2] It is extraordinary today to realize that one of his experiments was to show that the arteries contained not air but blood. He did it by tying an artery in two places and making an opening in between. How is it possible that this was not always obvious? The answer is probably that after death the arteries contract and empty themselves but the veins remain full of blood. It would be rare to *see* an artery in life, without deliberate vivisection, since it retracts into a wound. That there was a pulse was well known, but what was pulsing, and in what direction, was another matter. It was not until the thirteenth century that it was shown that you could recognize arterial bleeding by the way it spurted. Indeed, blood itself would be thought of quite differently: it was obviously one of the body fluids, but that it circulated was unknown (the very idea of a pump lay far ahead), as was the need of that continual circulation for the health of the whole body. It would not be clear that loss of blood was especially undesirable, any more than loss of urine, or stools, or vomit; in fact to think of some undesirable humour being carried away in blood-letting makes some sense of that practice. So it is equally not surprising that the tourniquet to stop blood loss also lay far away in the future.

The anatomy of the heart was, of course, long known in a very rough way. But it is remarkable that a treatise of the Hippocratic corpus on the heart, written in about 280 BC, that quite correctly describes the valves of the heart as the root of what we would now call the pulmonary artery and the aorta, and their function in making it impossible to blow air or pour water into the chambers of the heart through these vessels, adds 'especially on the left, for that side has been constructed more precisely, as it should be, since the intelligence of man lies in the left cavity'. Galen helped to counter this, in demonstrating the control of the body by the brain, by discovering that in a live animal, cutting a particular nerve in the neck known to come from the lower brain abolished movement of the larynx: this showed a control of speech by the brain, as well as making clear that nerves in some way controlled muscles. In a dead body, nerves and tendons look quite alike: both the Greek word 'neuron' and the Latin 'nervus' can mean either nerve or tendon. Only with experiment in life could the distinction become obvious between a tendon, essentially a tough cord transmitting mechanical force, and a nerve, which carries an activity that causes muscles to contract in response to reflex or volition. So we would not have known that the pinky-grey pulp in our skulls was what we now call a central nervous system, the 'seat' of intelligence, consciousness, learning, memory, control of voluntary action, mood, and emotion. Our thoughts, our vocabulary, and our outlook on the world, on ourselves, and on animals, would be profoundly different. The very knowledge of our relationships to the animal kingdom and of the evolutionary connection, and our understanding of their needs and behaviour that in part motivates the modern animal welfare movement, would be lacking. If we reflect on how unbalanced our views would be, we can say, indeed, that we *dare* not cease trying to understand how our bodies and those of animals function: we dare not let our knowledge of the inanimate world outstrip our knowledge of its inhabitants.

Practical benefits

This is a huge subject. Figure 1 provides a summary picture of part of it. The growth of animal experiment is shown here since 1890, using a logarithmic scale, so that it describes the proportionate increase in successive decades. Beneath it are given the names of particular substances or drugs, or sometimes classes of drugs, that have been important in human or veterinary medicine and that required animal experiment for their introduction. The approximate date of their introduction is opposite the first letter of the name concerned, so that we can see the successive 'ages' of the vaccine, the vitamins, the antibiotic, and some other contributors to the therapeutic revolution.

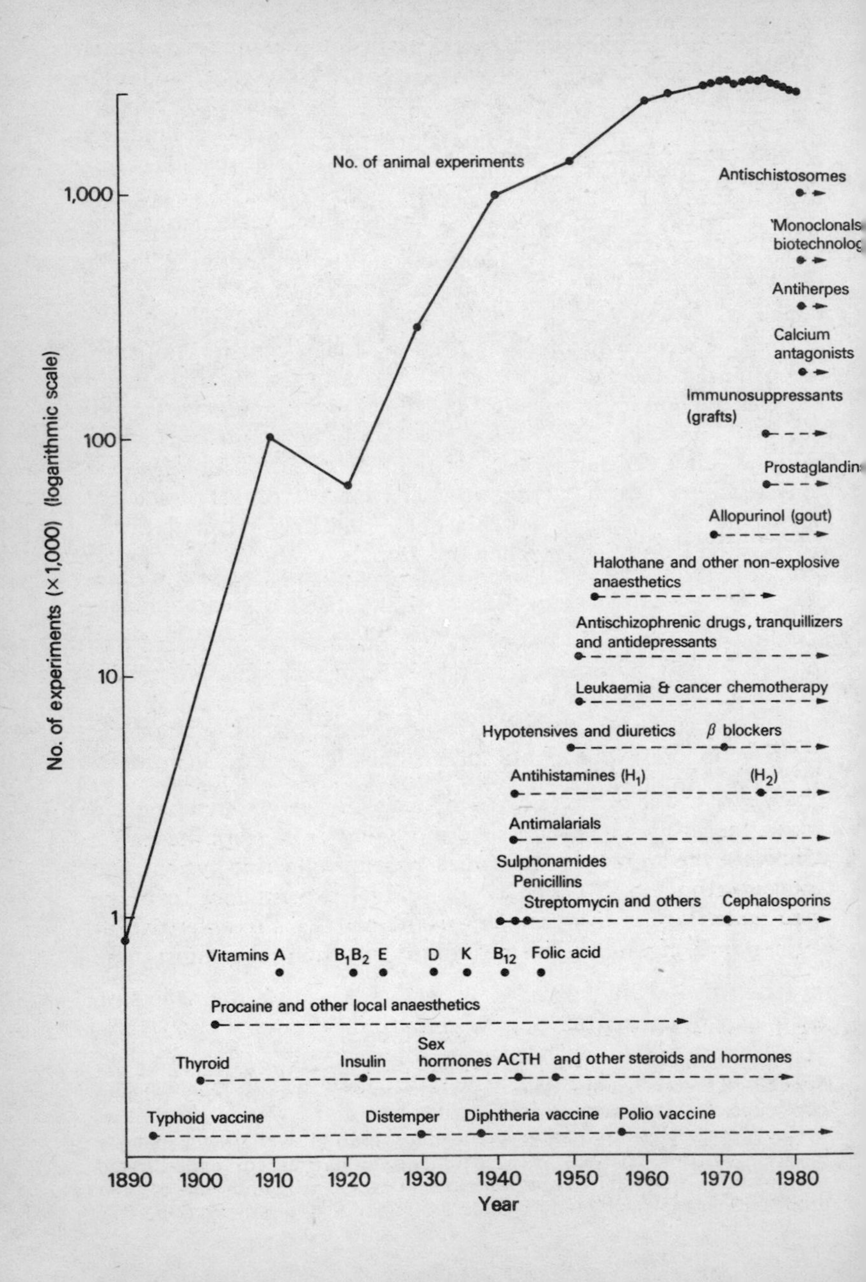
No. of animal experiments
No. of experiments (×1,000) (logarithmic scale)
1,000
100
10
1
Antischistosomes
'Monoclonals
biotechnolog
Antiherpes
Calcium
antagonists
Immunosuppressants
(grafts)
Prostaglandin
Allopurinol (gout)
Halothane and other non-explosive
anaesthetics
Antischizophrenic drugs, tranquillizers
and antidepressants
Leukaemia & cancer chemotherapy
Hypotensives and diuretics
β blockers
Antihistamines (H1)
(H2)
Antimalarials
Sulphonamides
Penicillins
Streptomycin and others
Cephalosporins
Vitamins A
B1B2 E
D
K
B12
Folic acid
Procaine and other local anaesthetics
Thyroid
Insulin
Sex
hormones ACTH
and other steroids and hormones
Typhoid vaccine
Distemper
Diphtheria vaccine
Polio vaccine
1890
1900
1910
1920
1930
1940
1950
1960
1970
1980
Year

Some caveats

Considerable caution, however, is needed in making such claims. Doubt about the fact of medical progress has been voiced, and some comments are very misleading. Thus the following remarks, attributed to Dr David Owen, have been claimed as grounds for scepticism: 'A man aged fifty in 1841, when reliable records began, could expect to live a further twenty years; by 1972–4 a man aged fifty could expect to live another twenty-three years. So, despite the improvements in health care in the intervening time, life expectancy had increased by just three years.'[3] One can well believe that in the 1840s, those tough enough to survive all the diseases which removed over half of the population before they were fifty might well survive a good while longer. But the chance of reaching the age of fifty was far lower than today. The expectation of life at birth was then of the order of forty-two years, some thirty years less than the expectation today, a very different figure from that quoted above. So, not only has life expectancy increased, but also the quality of that life has been improved because so many disabilities have been removed. The only real interest in the quotation is in the question of how long we wish to seek to live, and in the intriguing scientific question of what biological process it is that actually limits the duration of our lives. Our attitude to the ideal length of life perceptibly changes as our age approaches the critical zone!

There has also been a good deal of recent criticism of medicine, suggesting that it has not been therapeutic discoveries, but rather improvements in nutrition, or hygiene, or preventive medicine, or changes in the characteristics of diseases themselves, that have led to the longer life and diminished morbidity of Western life today. Of course the knowledge that made possible effective hygiene and good nutrition itself depended on animal experiment. To make sure water is free from cholera, salmonella (gastro-enteritis), or legionnaires' disease organisms, or milk free from tuberculosis, one

Fig. 1. Animal experiment and medical advance

The first letter of the word or phrase describing the advance is set opposite the approximate date of its introduction. Broken lines indicate continued development.

The growth of animal experiment since 1890 is also shown using a logarithmic (proportional) scale of the total number of experiments performed in the UK. The numbers are plotted at wide intervals of time until 1970, after which yearly figures are given. See also Table 8 (p.109).

must already know that such organisms cause disease, and how to identify them. All this knowledge stems back to the work of Robert Koch and others in the last century, working out how to determine whether a particular organism was harmful or harmless, in particular by showing in animals that a specific disease was *caused* by a specific organism: to this day organisms are found in the body that are merely chance contaminants. Equally, in nutrition, it required careful animal experiment first to *prove* that there was something in fresh vegetables and citrus fruits, or in fat, or in rice husks, that was essential for human health. The work of James Lind and others, pointing to the importance of fresh fruit in preventing scurvy, was vitally important; but it was not until experiment made it possible to identify vitamin C as such that clear nutritional advice and implementation was achieved. Similar experiment has been needed to identify other vitamins so that they could be synthesized for therapeutic use, and the amount in different foods determined. Nevertheless, if we trace the figures for growth of population and personal income over the centuries, we can readily recognize how mortality increases as pressure of population exceeds the availability of food, housing, and clean water, and how it improves as personal resources increase. It is no purpose of this book to argue that all advances in human health have stemmed from medical discovery based on animal experiment. The argument is much more important. It is that human and animal health depend on many things, including scientific discovery; that scientific work includes many branches, including animal experiment; and that it is a tragic error to try to set these various activities against each other.

When we seek to assess the part played by animal experiment, then, we must allow for these other factors. Equally we need to be critical before agreeing that a real benefit has been achieved. It is not enough, for instance, to accept dogmatic medical statements that this or that drug is valuable. The most vivid illustration of this I know is the opinion of J. A. Paris, President of the Royal College of Physicians from 1844 to 1856, that calomel, soon to be condemned and now abandoned, was outstanding as 'a preparation more extensively and more usefully employed than almost any other article in the whole range of the materia medica'.[4] The tracing of the rise and fall of gold therapy for tuberculosis between 1924 and 1944 by D'Arcy Hart provides another example.[5] Today we recognize explicitly the way the course of a disease may fluctuate. In the past, the chance taking of a remedy just before a remission seemed conclusive evidence of curative effect: we are more cautious now. We are equally critical of patients' reports, now that we know of the 'placebo response'. In this, with almost any disease assessed chiefly by patients' testimony (common cold, 'rheumatism', indigestion), apparent relief in 30–50 per cent of cases can be obtained, for a while, with distilled water or sugar. We can understand the placebo response better, too, with the recognition that the report of pain or suffering depends on two things: the actual discrimination of a painful stimulus, and *the liability to report it* as pain—a liability that is very susceptible to other influences. But our techniques of clinical trial enable us to distinguish the reliable from the transitory.

A last difficulty in finding cogent evidence of practical benefits is that we are largely restricted to statistics on *mortality*. The measurement of *morbidity* (suffering, illness, disability, handicap) in quantitative terms is extremely difficult, and reliable large scale data are lacking. So we will largely concentrate on mortality figures. That means we will have to leave out of account here any relief given to the sufferer from hay fever, the insomniac, the arthritic, the epileptic, the itching, the disfigured. The reader may, however, be able to fill in with his own experience.

Benefit to man

Our first procedure, then, is to look for mortality or other general statistics, taking careful account of previous trends and other factors. We shall see some abrupt changes in these statistics, but the evidence will not rest on that alone. There will also be the knowledge, for instance about the sulphonamides and antibiotics, that the drug concerned is known experimentally to be effective, that it was given in adequate amounts, and that individual case histories exist that convincingly illustrate the benefit. When general statistics are available, they serve to show that the effect is substantial at a population level. We may begin with some examples from bacterial disease.

First, **puerperal sepsis**. Figure 2 shows the statistics for maternal deaths and puerperal sepsis (childbed fever due to streptococcal infection of the genital tract), with a dramatic fall beginning about 1935 when sulphonamides were introduced, to be followed in 1940 by penicillin. As a result maternal mortality due to puerperal sepsis fell from around 200 per 100,000 births up until 1935, to 70—one-third of that figure—by 1940, and only about 5 in the 1960s.[6]

Second, **lobar pneumonia**. Pneumonia has a variety of causes,

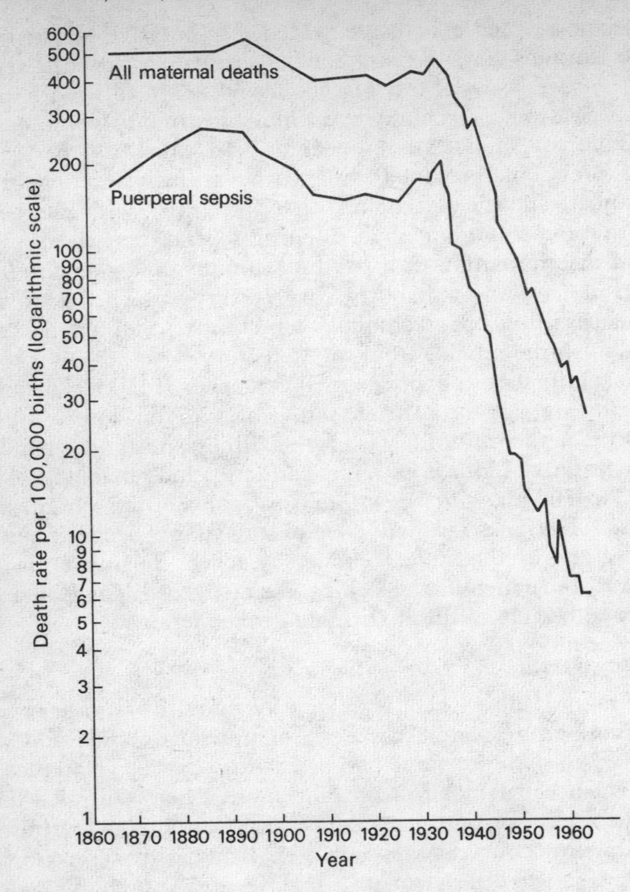

Fig. 2. Maternal death rate per 100,000 births in England and Wales, 1860–1964

Ten-year averages are given between 1861 and 1890, five-year averages between 1891 and 1930, and annual rates between 1931 and 1964.

Derived from the Registrar General's Decennial Supplement, England and Wales, 1931, Part III, and the Registrar General's Statistical Review of England and Wales, various years. From the Office of Health Economics (1966).

including viruses, whose treatment is still inadequate. But Figure 3, giving deaths specifically from lobar pneumonia (a very characteristic consolidation of the lung caused by the pneumococcus) in

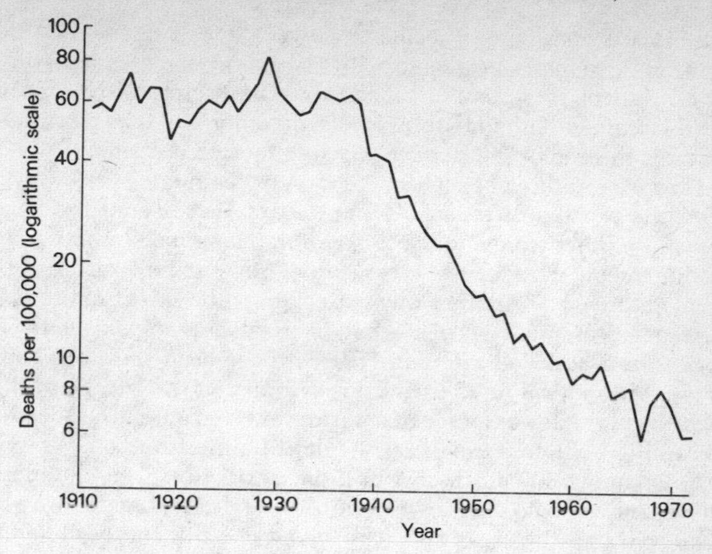

Fig. 3. Death rate from lobar pneumonia in middle-aged men (45–64) in England and Wales, 1911–72
From Anderson (1977).

middle-aged men, shows the abrupt fall from around 60 per 100,000 to about 6 in 1970.[7]

Third, **rheumatic fever**. The full results of bacterial infection do not always show themselves during the acute stage of the illness. An important cause of heart disease in the younger age groups used to be infection by a particular strain of streptococcus, for instance by an attack of tonsillitis or scarlet fever. Some years later, in a proportion of cases, this could result in attacks of rheumatic fever or chorea (St. Vitus's dance) associated with progressive heart disease. Although heart disease from other causes has continued to increase steadily until very recently, one of the effects of the introduction of chemotherapy was a reduction of heart disease at younger ages; thus deaths between the ages of fifteen and twenty-four fall from around 200 per million in 1935 to a tenth of that thirty years later.[6] Also there was a corresponding reduction in the occurrence of chronic disability in those who did not die.

Next, let us consider some examples of vaccines and antitoxins. First **diphtheria**, for which first an antitoxin was discovered to treat the disease, and then a vaccine to prevent it. Figure 4 shows the data, and also raises a valuable point about how one uses statistics. The result of these two advances was the decline of childhood deaths from diphtheria from levels of around 800 per million towards the end of the last century to almost zero today.

Now we can think of such a reduction in death rate in two ways, which is worth a digression. A natural way is to count the lives saved—in other words, to use an *arithmetical* scale. But that says nothing about how many lives were at risk, or how many lives were *not* saved. So another way is to express the saving of life as a percentage of those at risk—that is, to use a *proportionate* or *logarithmic* scale. The data in Figure 4 have been shown in both ways, arithmetical on the right, logarithmic on the left. From the right-hand graph we can see that antitoxin saved around 600 deaths per million children each year, while immunization 'only' saved the remaining 300 deaths per million: antitoxin seems the more important advance. But if we turn to the left-hand graph, we see that even after the use of antitoxin, about one-third of the children still died, while immunization was almost 100 per cent successful; so now vaccination seems the more important.

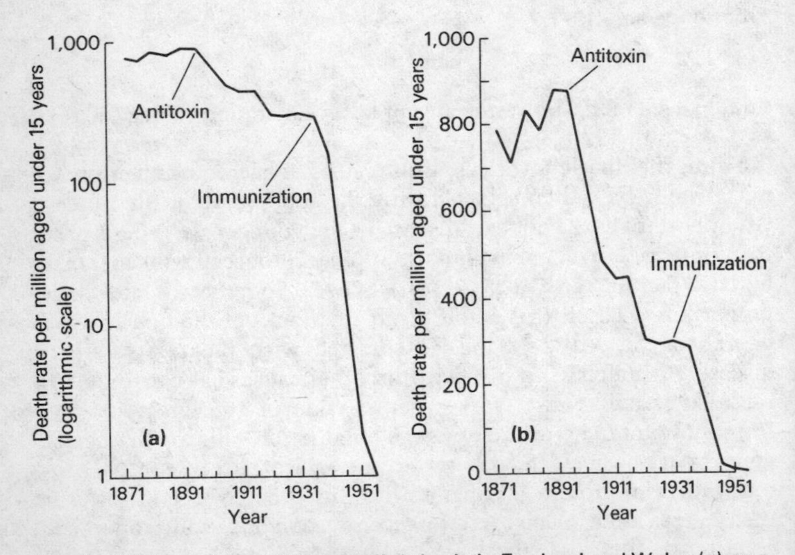

Fig. 4. Childhood mortality from diphtheria in England and Wales (a) on a logarithmic scale and (b) on a linear scale
From Altman (1980).

These contrasting approaches occur quite commonly. For instance, a taxpayer may prefer to think in terms either of the number of pounds, or of the proportion of his income, that he pays. Or with a rare cancer we may either think of the total number of times it occurred, or express that number as a percentage of all cases of cancer. There is no absolute reason for preferring one way of presenting the data to the other; it depends on the question at issue. If we need to know, for instance, the total lives likely to be saved, for whom provision needs to be made, the arithmetical approach is appropriate. But if we are thinking of lives that were *not* saved, of the task that still remains and where to direct our efforts, then we may well think in proportionate or percentage terms. In the present instance, the treatments are clearly effective by either approach. But the discussion may provide a gentle reminder to consider *how* such data are expressed.

Poliomyelitis is another important case.[8] Figure 5 is interesting in illustrating another point in the handling of data. One might, for instance, simply have taken, as baseline, data from some date (say

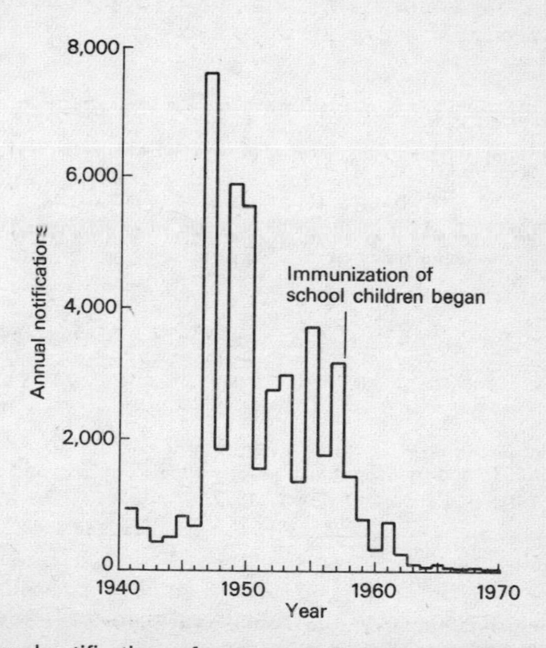

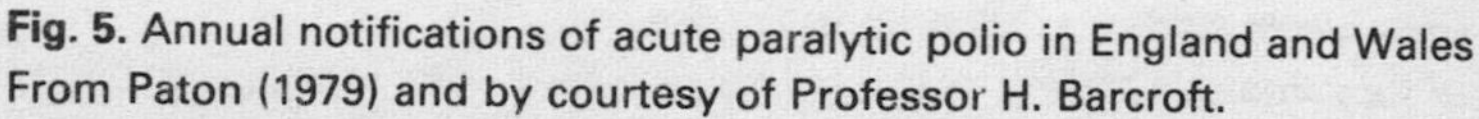
Fig. 5. Annual notifications of acute paralytic polio in England and Wales From Paton (1979) and by courtesy of Professor H. Barcroft.

1946 or 1955) during one of the epidemics, so as to make the improvement as dramatic as possible. But a wider look is more instructive: for, if we move the baseline back to a period between epidemics, we see that the introduction of polio vaccines has virtually removed not only the epidemics but also the background incidence from which the epidemics grew. Poliomyelitis is a significant case in another way since it is very clearly not one of the diseases of which the incidence is reduced simply by raising the standard of living.

Measles is another childhood infection preventable by vaccination. Figure 6 from a Medical Research Council Report gives a vivid picture of how the introduction of vaccination in 1968, by building up a good proportion of protected children within a year, almost abolished the biennial epidemics as well as greatly reducing the background incidence.[9] The reduction in measles also meant a fall in the childhood bronchopneumonia to which it readily gave rise.

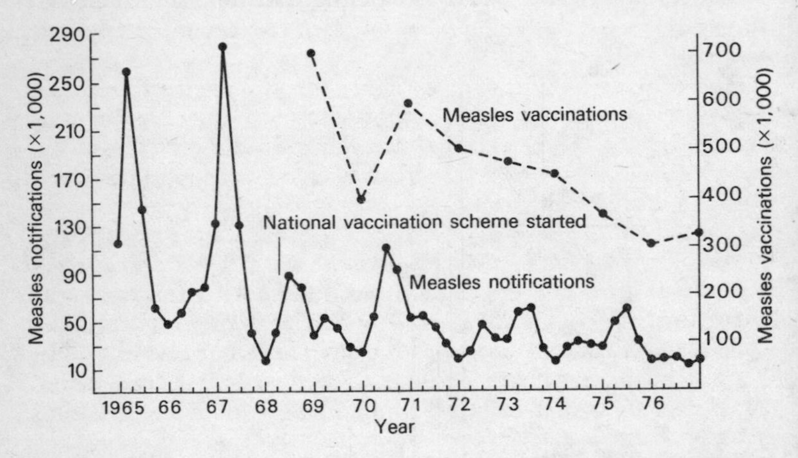

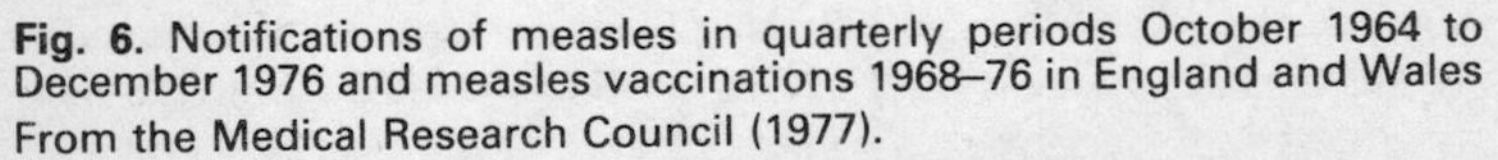
Fig. 6. Notifications of measles in quarterly periods October 1964 to December 1976 and measles vaccinations 1968–76 in England and Wales From the Medical Research Council (1977).

Whooping cough (pertussis) vaccine has been the subject of considerable debate. This is no surprise to the pharmacologist, who has used the toxicity of the killed organism for decades as an experimental tool.

The damage that a micro-organism does can arise in either or both of two ways. It may produce and release a poison into the blood—an 'exotoxin'; diphtheria is an example, and that is why the antitoxin was so successful. Alternatively, the body of the micro-organism, after it has multiplied and then died, may be poisonous (an 'endotoxin'); the whooping cough organism is one of these. Now the body of the micro-organism also contains the materials that produce immunity to the disease, and vaccines were made originally by simply injecting preparations made from dead organisms. If they are themselves poisonous, it obviously becomes necessary to find ways of separating the endotoxin from the constituent that produces immunity; and this offered some difficulty with the pertussis organism. In the process, it was necessary, of course, to have some measure of the effect of the endotoxin, so that one could be sure that it had been got rid of as far as possible; and it was from such work that its properties came to light. The whole story provides a reminder that a vaccine against some infectious organism is not, so to speak, automatically available; one must separate the toxin properly from the antigen; and until one can do this, vaccination itself may incur the danger of disease.

For pertussis the difficulty of doing this threatened to create a dilemma calling for the wisdom of Solomon; for it seemed that whooping cough vaccination carried a significant risk of causing damage to the brain. It is of course true that whooping cough itself carries risks. In a six-month study during the 1974–5 epidemic ten children died and one child in ten was ill enough to be admitted to hospital. In the 1977–9 epidemic, of those admitted 12 per cent had pneumonia and 5 per cent convulsions.[10] But despite this there was a real dilemma; for an unvaccinated child might not get whooping cough, and if it did it would not be any *particular* person's fault. But vaccine brain damage would have been a result of the parents' and doctor's decision, and they would find it hard to feel free from blame.

In the event, however, it appears that the risk of nervous damage has been much overestimated, chiefly perhaps because it has been too readily assumed that *any* signs of some effect on the brain after vaccination (usually a multiple vaccination) can be due *only* to the whooping cough component, and that there was no intervening disease of any sort. After a recent seven-year study of adverse effects of vaccination by the Public Health Laboratory Service Epidemiological Research Laboratory, involving 400,000 injections of vaccine, there was no evidence of brain damage due to it.[11] As the report says, 'If this syndrome exists, instances would probably have been discovered.' The acid comment by 'Minerva', in the British Medical Journal, seems justified: 'So what has the national press, radio and television had to say about this reassuring important study? Not a word; or if anything has appeared Minerva missed it.'[12]

Figure 7a, from another recent paper, gives an interesting picture of the incidence of cases in relation to the records of vaccination: since the number of cases must depend on the number of children born, the annual births are included, from which the proportion vaccinated can be estimated.[13] The fall in whooping cough is seen from 1950 onwards, as the proportion of vaccinated rose, followed by the recent resurgence after near freedom for ten years when a media campaign on the dangers of vaccination led to a drop in its use. This is a reminder that, until the organism can be completely eliminated, the need for a high uptake of vaccination remains. One can also say, after reviewing the evidence, that in principle whooping cough epidemics could be eliminated: Figure 7b illustrates this for Fiji.

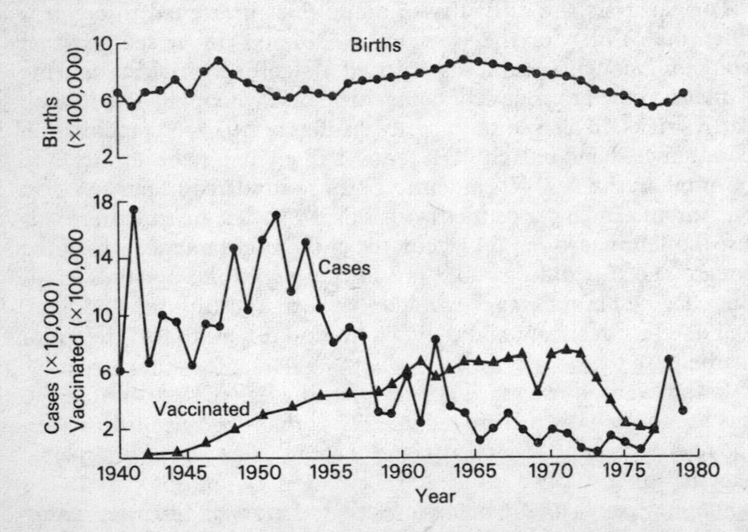

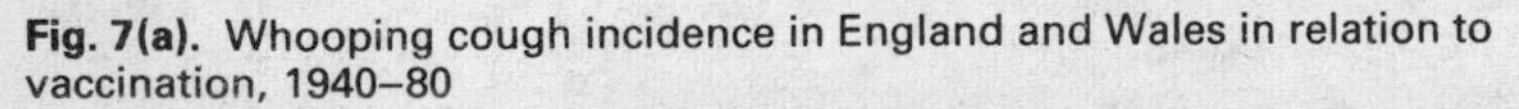

Fig. 7(a). Whooping cough incidence in England and Wales in relation to vaccination, 1940–80

The annual births are included since this influences the number of cases. From Anderson and May (1982).

The one great example of the elimination of a disease is, of course, **smallpox**, and it needs no illustration. But it is worth recalling that, for instance, death rates in India as a whole used to range up to 2,000 per million in a year, with far higher local incidences;[14] and that in those who did not die, the scarring for life with pockmarks could be a devastating disfigurement. It might be asked how far the use of animals contributed to this success, when one remembers that Edward Jenner's famous experiments on vaccination, recorded in 1798 in his *Inquiry into the causes and effects of variolae vaccinae*, were all on human subjects. For a long time, the principal use was as a standard source of cow-pox lymph, raised on the abdominal skin of calves. But the abolition had to await the ability to identify viruses, and for modern immunological knowledge, for both of which animal experiment was necessary in a variety of ways.

A lesser, but interesting, case is that of **dental caries**. There is an organism particularly associated with caries, named *Streptococcus mutans*. Monkeys fed on a high-sugar diet develop caries much as we do. It has been found, first, that a crude vaccine prepared from

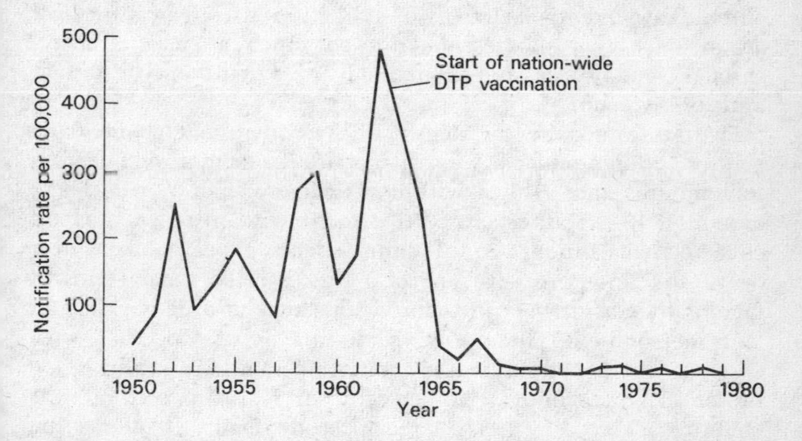

Fig. 7(b). Whooping cough notification rates per 100,000 population in Fiji, 1950–80

Vaccination acceptance rate in Fiji between 1972 and 1980 was about 85 per cent, calculated by subtracting infant deaths (5,782) from live births (162,430) and dividing the result into the number of completed courses of vaccine given to pre-school children (132,817). Vaccination was with a triple vaccine for diphtheria, tetanus, and pertussis (DTP). From Pollard (1983).

the organism could reduce this caries incidence considerably, and then that a purified antigen has the same ability.[15] One can only speculate whether there will ever be a case for widespread immunization against caries. But the work has thrown considerable light on a condition that causes much trouble and expense; and it emphasizes both the importance of oral hygiene, and that general health and the capacity to generate an immunological response may play a role in protection against caries.

A more important area is the group of **chronic cardiovascular diseases** (Figure 8). One must expect chronic diseases to respond to therapy only gradually. But it is worth noticing that with the advent around 1950 of drug therapy for hypertensive disease—a treatment that was clumsy to begin with but has got better and better over the years—the steadily rising death rate from hypertensive disease has stopped rising and now is steadily falling, together perhaps with the death rate from strokes, to which hypertension may lead.[16] In America there are now signs that death from coronary heart disease is also beginning to fall.

Finally **childhood leukaemia**, usually the acute lymphocytic form. (In the older or middle-aged person it is commonly the chronic myeloid type.) Here I would like to illustrate how medical advance is often progressive—not so much a sudden 'breakthrough', but a succession of small advances that finally results in a major dividend.

Figure 9 records the first steps.[17] With no treatment, about 50 per cent of children under fifteen with acute leukaemia were dead in four months and all died within little more than a year. Then, around 1946–8, rather savage treatment with nitrogen mustard added perhaps a month to this time. Then some early anti-cancer drugs, that interfere with one of a cancer cell's natural growth factors (folic acid) were introduced (this type of drug is called an 'anti-metabolite'). These folic acid antagonists, together with steroids, lengthened survival to about nine months; and with some newer anti-metabolites a 50 per cent survival of twelve months could be achieved by 1953. A recent study (Figure 10) shows the next step. Here the data are expressed in a different way, using not death rate but the presence or absence of leukaemia cells in the blood as an indicator (a 'haematological remission' is when they disappear after treatment). The upper figure shows that such a remission would now last for twelve months; since death would not follow till some time later, this represents further improvement on Figure 9. But the lower figure shows the result of an

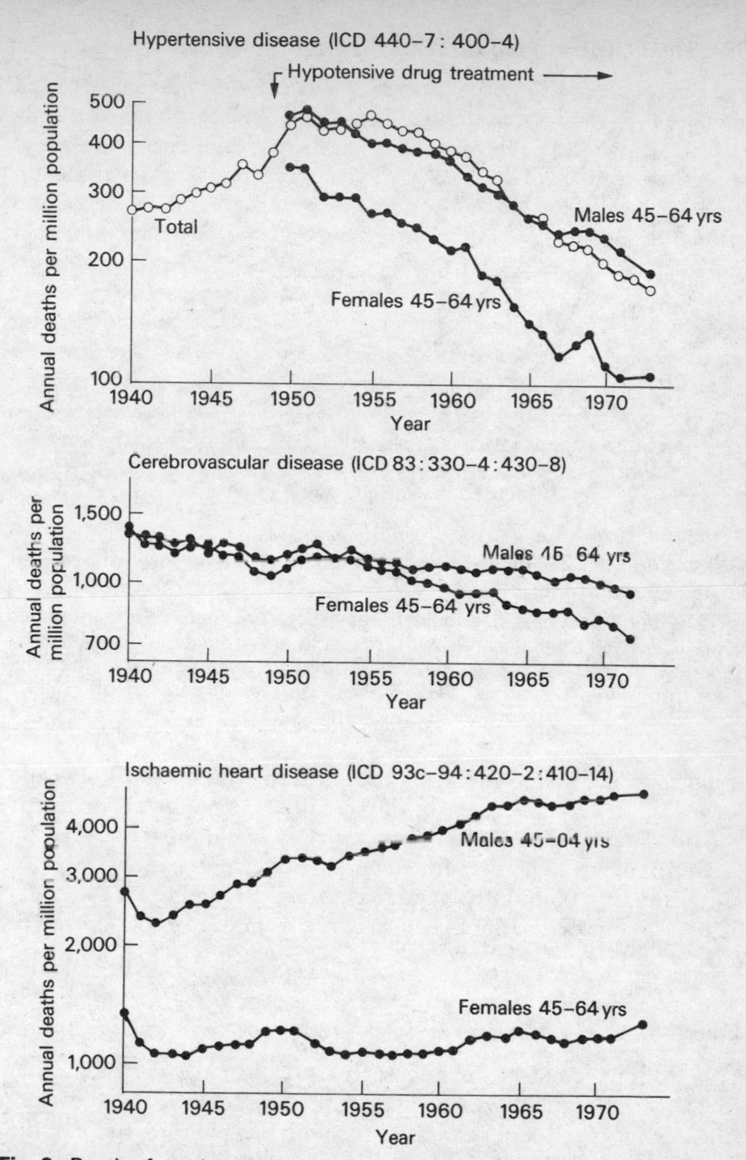

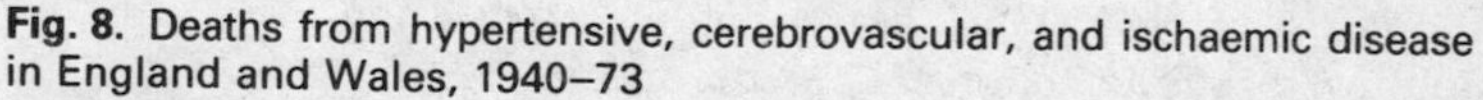

Fig. 8. Deaths from hypertensive, cerebrovascular, and ischaemic disease in England and Wales, 1940–73

The International Classification of Diseases (ICD) was revised 1949–50, 1957–8, and 1967–8, and the ICD categories used are shown. Death rates, adjusted for change in the age and sex distribution in the population, have been extracted for males and females of 45–64 years of age.

Data are from the Registrar General's Statistical Reviews of England and Wales. From Paton *et al.* (1978).

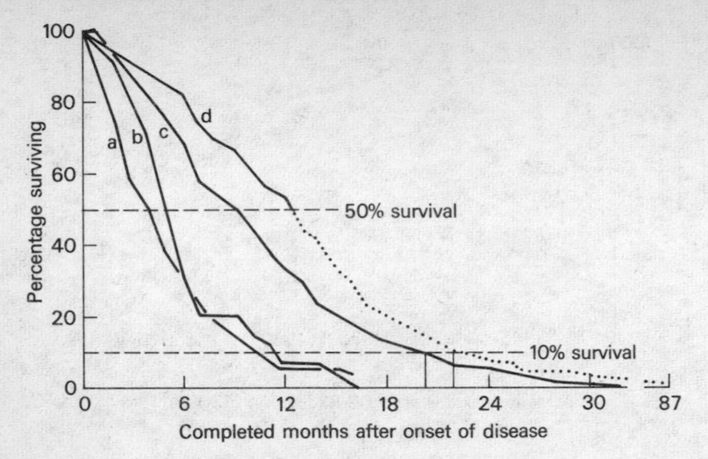

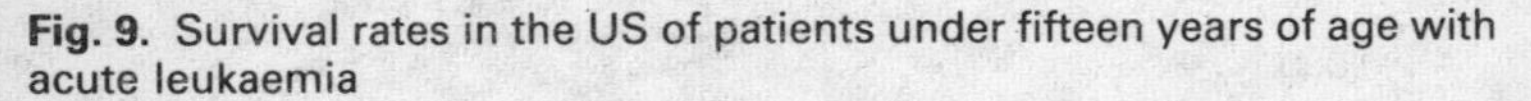

Fig. 9. Survival rates in the US of patients under fifteen years of age with acute leukaemia

(a) 218 untreated cases; (b) 34 cases treated with nitrogen mustard, 1946–8; (c) 160 cases treated with folic acid antagonists and/or steroids, 1948–52; (d) 184 cases treated with folic acid antagonists, steroids, mercaptopurine, azaserine, 1952–5. From Burchenal and Krakoff (1956).

important new step, irradiating the skull to attack leukaemia cells lodged in the brain; because of the so-called 'blood–brain barrier', it is hard to reach them with drugs, yet they could initiate a relapse. With this additional procedure, over 60 per cent of patients were in remission for 100 months; and it is thought today that more than 50 per cent of such patients can be cured if they are under the age of ten.

Cancer

This leads us to one of the largest fields of animal experiment, that of cancer research, which deserves more detailed discussion. Although it may seem pedantic, it is worth explaining some of the words used. 'Neoplasm', as its English translation 'new growth' implies, means an abnormal growth of cells. Such a growth may be termed 'benign', like a wart, or 'malignant' if it is locally or generally invasive, when it is also called a 'cancer'. The word

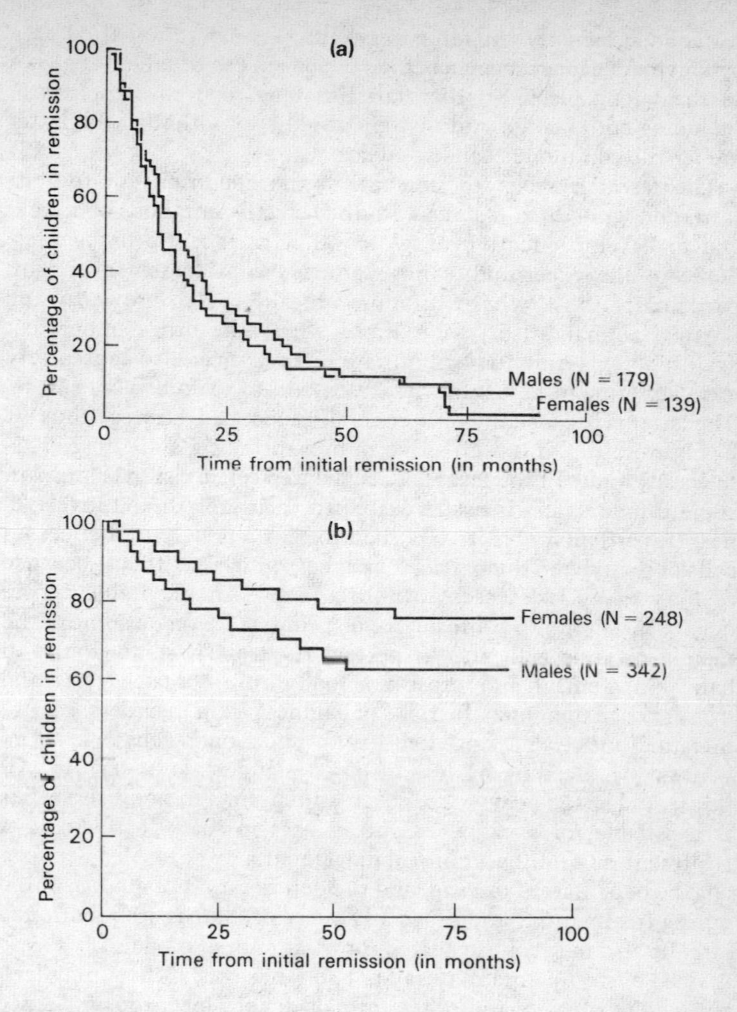

Fig. 10. Haematological remission in boys and girls in the US under 16 years of age with leukaemia

In the first study (a) various methods of chemotherapy were used without any special treatment of the brain. The second study (b) was designed specifically to test the further protection achieved by irradiating the brain. N is the number of children treated. From Sather *et al.* (1981).

'tumour', originally meaning a swelling, is also often used as a synonym for cancer. Even a benign neoplasm can be fatal if it grows in the wrong place; so the statistics may refer to 'neoplasms', including both benign and malignant growths, although the latter are far the commoner cause of death.

The research seeks to understand the conditions of tumour initiation, growth, and spread, to find selective methods of attack, and to develop effective drugs or other therapeutic procedures. Some of the experiments have aroused considerable criticism, particularly those where tumours are allowed to grow causing obvious animal misery, and those where the drugs under test against an experimental tumour are toxic. Cancer also causes very great suffering in man, and the treatment, too, can be very severe. The latter is because the only generally successful strategy thus far has been to find drugs effective in preventing cells dividing. This takes advantage of the fact that accelerated cell division is found in some tumours, and is associated with their growth and invasiveness. A particular objective is then to seek a preferential effect on cells of the type of the particular cancer, which is difficult because of their many likenesses to normal cells. An inevitable consequence is liability to some degree of a similar action on some of the rapidly dividing cells in our normal tissues. These are found in hair, skin, stomach and intestine, blood-forming organs, testis, and ovary; hence there may be risks of baldness, skin disorders, gastro-intestinal upset, anaemia, liability to infection because of loss of white cells, bleeding by loss of blood platelets, and sterility. All this would be too heavy a cost of treatment for a minor disease, but is justifiable for a cancer. Cancers vary in the effectiveness of treatment, so a difficult clinical judgement may be needed between chance of palliation or cure, and the deterioration of quality of life during treatment. Some especially successful areas are mentioned later. It also raises, for research, how far animal models of disease may be pushed. Some may say not so far as this. Others may feel, since domestic pets and other animals die of cancer too, and since humans currently experience anything that animals are exposed to, that it is legitimate, provided (as we shall review later) that the science is sound and the suffering reduced to what is unavoidable.

Before considering the progress achieved, we must face the claim that is sometimes made, that so far from any success in reducing cancer mortality, it has substantially increased despite all the research. This has prompted some to call for such research to be stopped. The point is an old one: Table 1 shows the figures given in

Table 1. Annual deaths from cancer per million population in England and Wales, 1871–1904

1871–75	1876–80	1881–85	1886–90	1891–95	1896–1900	1901–04
445	493	547	631	711	800	861

an article in 1911 for crude annual death rates from cancer per million people in England and Wales.[18] But the shrewd comment was made that it was not clear how far this doubling in overall death rate represented a real increase in cancer incidence and mortality, how far it was due to more accurate diagnosis and improved recording, or how far it was simply due to the increased length of life, so that a larger proportion of people reached the age at which cancer is most frequent. Today such crude figures for death rate are no longer used, but only figures adjusted to a form not materially affected by the age distribution of the population as a whole. The usual form is that of *'age-specific' death rates* for particular types of cancer (i.e. rates in some narrow range of age—useful in detailed study of a particular cancer, but giving an overwhelming mass of data if applied to a population over a wide range of cancers); and *'age-standardized' rates* (where the rates are recalculated, knowing the age distribution of the population from which they derive, to the rates that would be found in some standard population; for instance, Doll and Peto have used the US 1970 census figures as the best available standard[19]). Once the changes in age distribution due to changes in birth rate and longevity are pointed out, the fallacies in using crude death rates are obvious, but their magnitude will be less familiar. A few examples may be quoted from Doll and Peto. They describe some US changes in age distribution between 1953–7 and 1973–7: for white males, the population of age 0–4 declined by 18 per cent; between 20 and 24 rose by 72 per cent (the result of a surge in the birth rate); between 40 and 44 was almost unchanged; between 60 and 64 rose by 26 per cent; and between 80 and 84 rose by 78 per cent. There were similar changes for white females, with a rise in those over 85 by 225 per cent. Non-whites show general rises at all ages, but also the same increase in the older groups. To such causes of misunderstanding must be added allowance for errors or changes in diagnosis, registration, and sampling, each a matter of considerable study. The most reliable statistics date from 1950; with care, data may be used back to 1933; before that the uncertainties are too great. All this produces a greater caution in dogmatic statement, the more expert the investigator.

Some conclusions, despite these difficulties, are possible.[19] First, on the general trend of age-standardized cancer mortality: if one sets aside cancers related to smoking (in the respiratory and upper alimentary tract), the trends in mortality over the most recent ten-year period for some 30 groups of cancers in both sexes are mostly downwards, by amounts generally of the order of 1–3 per cent a year. The chief exceptions are pancreatic cancer in women and melanoma in both sexes.

Second, to know if this is due to improvement in cure rate, it is necessary to have information about cancer incidence. Here the statistical data are inevitably subject to even greater difficulties of interpretation. But a cautious interpretation of the evidence does point to a general small improvement in 'relative survival rates' (a relative survival of 100 per cent would indicate a risk of death the same as that of the standard US population as a whole matched for age and sex): thus the relative survival rate has risen from 60 per cent in 1950–54 to 68 per cent in 1970–3 for cancer of the female breast, and there are similar changes with all the common cancers except stomach and pancreas and lung. There is greater success in three areas:

(a) **Hodgkin's disease**, where survival over the same periods rose from 28 per cent and 34 per cent to 66 per cent and 69 per cent for males and females respectively.

(b) **Leukaemias** in the young, where three-year relative survival had risen from 5 per cent in 1950–9 to 24 per cent in 1967–73 for those under thirty-five years of age. Figures 9 and 10 show reason for believing that population statistics for childhood leukaemia today may show a still greater hope of survival.

(c) The less common **cancer of the testis** and **choriocarcinoma** have also shown major improvement in survival rate.

Lung cancer requires special discussion, as the outstanding exception to the general drift downwards in cancer death rates. The figures in Table 2 make the point.[20]

It is now generally accepted both that the association between cigarette smoking and lung cancer is real, and that it is causal. In what way, if any, has animal experiment contributed to the recognition of this cause, to the understanding of the mechanism involved, or to the prospect of reducing the mortality? The recognition of a causal association must certainly be attributed

Table 2. The growing contribution by lung cancer to total deaths from neoplasms: deaths per 100 million US population under 65 years of age

	1935	1955	1978
Males			
All neoplasms except respiratory	58,709	57,176	52,538
Respiratory cancer	5,812	19,495	33,816
Females			
All neoplasms except respiratory	90,108	71,893	58,618
Respiratory cancer	2,102	2,820	12,064

primarily to skilful and tenacious research by epidemiologists, studying the incidence of disease in social groups or whole populations. But it seems that the acceptance of such a conclusion by the general public also requires some understanding of the mechanism involved. This can be illustrated by the history of research into sepsis. The Hungarian investigator Semmelweis had proved by around 1850 that childbed fever was produced by the transmission of infective material by accoucheurs into the maternal genital tract from the post-mortem room or from other patients. Yet no one believed him, and he died a disappointed man. Just twenty years later Lister's antiseptic method was winning international recognition. The main reason why Lister's conclusions were accepted when Semmelweis's had been rejected was the effective birth of bacteriology in the intervening period. No longer was it necessary to postulate an unknown, hypothetical infective material; instead there were microscopically visible organisms. The mechanism linking cigarette smoking and lung cancer is still by no means clear. But our understanding has progressed considerably. Examples are the explicit demonstration that cigarette tar painted on a mouse's skin can cause cancer; the chemical identification of highly potent chemical carcinogens in tars (the benzpyrene series); the recognition of the complexity of carcinogenesis and of the processes of 'initiation' and 'promotion'; the discoveries of how these seemingly inert compounds are chemically transformed in the body to much more reactive substances; and the identification of a chemical reaction between these substances and the constituents of the cell's genetic material.

These are only a few of the developments, directly or indirectly making the pattern of the mechanism more intelligible.

There has been one further step. The animal work has indicated that it is not primarily the nicotine, nor the gases nor the carbon monoxide in cigarette smoke that is carcinogenic, but something in the tar, and there has been a great deal of research into its composition and properties. Over the last two decades, the level of tar in cigarettes has been steadily lowered. Much research has also gone into exploring tobacco substitutes which could lower the risks of smoking, by lowering the yield of the various harmful substances. The epidemiological results, just beginning to emerge, are most encouraging. The effect is most obvious, as one would expect, in younger members of the population, for whom the 'lower tar era' represents a larger part of their lifetime experience of smoking. Table 3 shows some of the results.

Table 3. Annual deaths per million from lung cancer in England and Wales, 1940–80

Age group	Men born *c.* 1910 observed 1940–60	Men born 1930–50 observed 1980	Percentage reduction
30–34	40	13	64
35–39	98	45	54
40–44	293	134	47
45–49	597	378	37

Sir Richard Doll comments on this result that it 'detracts in no way from the importance of education and other measures to discourage smoking altogether, which can eventually have a much greater effect on total mortality from a variety of cancers and other disease; but it is unrealistic to think that a habit that has become so commonplace and involves so many financial interests can be eliminated overnight. It is evident, however, that people are willing to switch to low tar cigarettes and if, as now appears, such a switch leads to a major reduction in the risk of disease, we need to do everything we can to bring it about.'[21] The recognition of the role of smoking must be above all credited to the epidemiologists; but for their arguments to be accepted, and for the low tar cigarette to be introduced and accepted, there have been many other essential contributions, including that of animal experiment.

Cancer due to occupational and environmental agents

There is now a formidable list of agents established as able to cause cancer in man. Among those arising in everyday life, as well as smoking, are aflatoxin (produced by mouldy food in the tropics), alcohol, chewing betel, tobacco, and lime, certain parasites (for example, bladder cancer from schistosomes), ultraviolet light, and hepatitis B virus infection. Industrial work may involve exposure to a carcinogen, for example to asbestos, benzene, cadmium, some factors in furniture or leather manufacturing, nickel, beta-naphthylamine and other aromatic amines, polycyclic hydrocarbons, and vinyl chloride. Some substances used in medical treatment or in self-medication may also be carcinogenic, such as oestrogens, phenacetin, steroids, immunosuppressants in organ grafting, and alkylating agents. Their initial recognition arose in various ways from case history, epidemiology, and animal work. But with them all, the establishment and acceptance of the cause, the understanding we have, and the means of control, rest on no single fact or discipline, but on co-ordinated work in almost all branches of biology and medicine.

The demands for control, together with the number of substances, are largely responsible for a great deal of industrial testing of new compounds. Mere enumeration has begun to give the impression that it is substances of this type that cause the bulk of human cancer. But it is an important conclusion, in Doll and Peto's remarkable and convincing book, that all these industrial compounds together probably account for no more than 10 per cent of human cancer deaths, while tobacco and dietary factors are responsible for two-thirds, and viruses for a substantial proportion more. This conclusion calls into question the extent and nature of the testing of new substances, so extensively required of industry at present. Certain simple policies must, however, clearly be rejected. To stop all testing, including animal tests, and to rely solely on epidemiology would result in thousands of avoidable deaths, and is clearly unacceptable. Public policy over the nitrosamines already illustrates the rejection of such an approach. Here are a group of chemicals that were found by animal experiment to be among the most potent carcinogens known in animals. As a result of this finding, and in advance of any epidemiological proof of a hazard in man, determined efforts are now made to reduce nitrosamine levels in food and elsewhere in the environment as far as possible.

Nor can one simply rely on non-animal methods of test alone

before human exposure is allowed. There are a variety of these: tests on enzymes, on the liability to mutate in bacterial culture (the Ames test), or on cultured mammalian cells. Alternatively, quick animal tests involving little suffering may allow recognition of an interaction with DNA, or of a genetic change, or of an effect recognized microscopically on reproductive cells. Such tests have the great advantage of speed, and of avoiding long-term animal exposure. Yet their predictive power is quite clearly not strong enough; and indeed it is unreasonable to expect such tests over a few days or weeks to predict what will happen over the whole of a human life-span. At the same time, long-term animal tests are also limited in predictive capacity, so that one cannot rely on them alone either.

There is evidently no simple answer. There is a lot to be said for Doll and Peto's suggestions for estimating which substances are most likely to matter: they propose using laboratory tests for what they call 'priority setting' rather than for formal 'risk assessment', and of then combining *in vitro* and animal tests together with epidemiological estimates of frequency and intensity of exposure. As a result of the current intensive debate on such matters, it is probable that the part played by animal experiment can be made considerably more efficient and economical.

Advances in surgery

The account so far given has focused chiefly on medical conditions, and on treatment with drugs and vaccines. This is both because of my own experience and because mortality statistics of the kind so far cited at the population level, are less readily available for surgical procedures. But we must nevertheless review some uses of animal experiment in surgery.

Once anaesthesia and the control of infection had opened the gate to advances in surgery, animal experiment began to find the best material to use for one of the surgeon's primary tasks—rejoining tissues that he has had to cut through. Experimental wounds were made under anaesthesia and then repaired with the test material. A wide range of **suture materials** has been tried: hemp, silk, catgut, various metals, tendon, horsehair, even strips of skin or blood vessels. The necessary qualities were gradually discovered. The suture material needs to be strong enough to hold, for instance, the abdominal wall together, even after exposure to the enzymes in the tissues. It needs to maintain its strength even in a moving organ such as the intestine, or when exposed to such fluids as gastric juice. It should be able ultimately to be reabsorbed, and yet maintain its strength until the repair mechanisms of the body have allowed scar tissue of sufficient strength to form. It must be sterilizable, non-irritant, and non-antigenic. The actual loads to which a suture may be subjected had to be measured. A special problem was that of how to suture, in a child, an organ that was still growing. One of the most delicate recent achievements is that of the microsurgery of blood vessels, with the development of materials and techniques that allow operation on blood vessels no more than 1 mm. in diameter. All this called for decades of experiment ranging from Lister's original studies in the last century up to the present day.[22]

A second major area lies within **orthopaedic surgery**. Before modern techniques could be developed, physiological experiment was necessary to discover, for instance, by using a dye (madder) that is taken up only by growing bone, that the long bones grow, not evenly along their length, but from their ends; that regeneration of bone occurs from the tough surface layer over the bone called the periosteum; and that the body can 'remodel' bone by combining bone removal with bone deposition.[23] Ways were found to graft bone, to ensure the sterility of the grafted materials, and to preserve such material in a hospital 'bone bank' for future use. Armed with such knowledge, it became possible to correct skeletal deformity, and, for instance, to hold back the growth of a normal limb in a child so that it should be in adult life the same length as a limb whose growth had been impaired.

A great field of advance has been in surgery of the hip joint. A common accident in the elderly is to break the neck of the femoral bone where it is set into the hip socket. A solution is to reset the fracture and then hammer a metal pin through the neck to support it. But some metals are liable, when surrounded by tissue fluids, to become electric batteries, and to set up electric currents which can dissolve bone and produce other changes that destroy the firmness of the support. So metals had to be found that were 'electrolytically inert', and that were compatible with the tissues. These metals could also then be used as splints to broken bones, so that some jockeys and racing motorcyclists are now liable to be walking ironmongery shops!

Surgery of the hip has gone still further, in patients with osteoarthritis, where the whole hip-joint is replaced with an artificial socket embedded in the hip-bone, into which fits an artificial femoral head embedded in the femur. Again, compatibility with the tissues had to be secured. For this the modern plastic has proved of great use; and the success of the whole research is shown by the waiting-lists for 'artificial hips'.

Another important area is **cardiovascular surgery**.[24] As Comroe and Dripps have shown in detail, this has been built on centuries of experimental and clinical work on both the normal function of heart and circulation, and on their disorders. Today an artery blocked by arteriosclerosis can be replaced or by-passed, either by a vein or by artificial materials, backed by knowledge of what happens to these, not just immediately after the operation, but over the following weeks and months. The **cardiac pacemaker** is another great advance, depending both on the knowledge of how to make an implant compatible with the tissues, and on a deep knowledge of the electric currents that flow when the heart beats, and of how the beat itself is initiated. **Cardiac surgery** only became possible (except in rare cases) with the development of two techniques: the heart-lung machine, that allowed the circulation to vital organs to be maintained while the heart was stopped; and deliberate hypothermia, where the body temperature is cooled so low and the need for oxygen by the brain and other organs is thereby so reduced, that an absence of blood-flow can be tolerated for the duration of an operation. Together with great surgical imagination and skill, this allowed the series of 'blue-baby' operations, where congenital abnormality of the heart threatens the life of an otherwise normal child. Along with this went the design and testing of artificial heart valves to replace those damaged by disease, and the working out of procedures for stopping a beating heart before operation, and of restarting it afterwards.

The heart-lung machine has a sister in the **artificial kidney**. Failure of the kidneys can arise in many ways: for instance from disease, from poisons or wrongly used drugs, or from traumatic shock such as accompanies gross crushing injury. The idea of 'renal dialysis'—that is of taking blood out of an artery, passing it over a thin membrane with a suitable salt solution the other side to let waste products diffuse away, and then returning it to the body—is about seventy years old. It only needs a glance at a microscopic section of a kidney to understand why it has taken so long to understand the physiological processes well enough to be able to make an artificial substitute. Eventual success has, in fact, created a dilemma. For while such artificial kidneys are satisfactory when kidney failure is only temporary, the cost of lifetime provision for many thousands of patients is very large indeed.

This has proved a great stimulus to **renal transplants**.[25] This, too, was envisaged by surgeons several decades ago; but it needed an understanding of immunology, of the physiological 'laws' which determine whether a grafted tissue or organ will be accepted or rejected by the body, before it became practicable. The problem was somewhat analogous to that of blood transfusion, for which a donor's blood group must be matched to that of the recipient; but the expression of individual characteristics in a whole organ is very much more complicated. The matching of what are called 'histocompatibility antigens' has greatly advanced. This, together with experimental development of ways of storing kidneys from dead people, and of drugs which can hold back immunological rejection (the so-called immunosuppressants) have opened the way to renal transplant as a general technique. If successful—and success rates are steadily rising—the operation constitutes in effect a cure, and the patient is freed both from the constraints and the costs of regular renal dialysis. As so often in medicine, the success has come from a succession of small steps, in improved immunology, better surgery, and better drugs, depending alike on clinical, animal, biochemical, and chemical experimental work. A similar account could be given of **corneal grafting**, first explored on rabbits' eyes. The same gradual improvement is visible with grafts of heart and lung, liver, pancreas, bone marrow, cartilage, and nervous tissue.

We could summarize what animal experiment, with other work, has offered to surgery by reviewing what would happen, say, to a soldier with gross injuries to the face from bomb blast. Blood loss could be compensated for by blood transfusion, and infection could be combatted by antibiotics. Surgical operation would be under a safe anaesthetic combined with analgesics if necessary. The suture material used would leave the minimum scarring. Skin or bone graft could help in reconstructing the face. If he could not swallow, artificial feeding of a carefully worked out nutrient 'cocktail' would be infused into a vein. If shock had produced renal failure, some form of dialysis could be used. The dressings applied would prevent infection, be non-irritant, prevent fluid loss, and foster healing. None of these procedures would have been possible without animal experiment.

The change in appearances of human disease

Our review of some of the historical benefits for which animal work was essential has been chiefly statistical. But to someone who was trained in medicine forty years ago, an equally cogent and more vivid argument is the simple abolition of some everyday manifestations of disease. One no longer sees infants with ears streaming pus, schoolboys with facial impetigo, beards growing from heavily infected skin, or faces pocked by smallpox or eroded by lupus, or heads and necks scarred from boils or suppurating glands. Drugs and a better diet have transformed the haggard patients with peptic ulcers. The languid, characteristically brown-skinned case of Addison's disease of the adrenals, the pale, listless patients of chronic iron deficiency or pernicious anaemia, and the cretin, or conversely the young woman with 'pop eyes' and overactive emotional behaviour—due respectively to thyroid deficiency or excess—are all being treated. The soggy hulk of a patient in the oedematous stage of chronic kidney disease is relieved by diuretics. As a result of polio vaccine and control of tuberculosis we see few crippled children: as one walks behind a group of youngsters today, varied as ever in shape and size, the marvel is how straight their limbs and backs are. The chronic arthritics with their sticks are being replaced by septuagenarians swinging along on their plastic hips. The patients now are rare that once one saw dying from an infected mastoid, struggling for breath in the last stages of heart failure, or dying from appendicitis, leukaemia, pneumonia, or bacterial endocarditis. For those without such memories, the pictures of Hieronymus Bosch, or the cartoons of Rowlandson or Hogarth, will give some inkling. Equally vivid, for those with some medical knowledge, are Professor Paul Beeson's accounts of how therapy has changed in the last half century, and of how these changes were brought about.[26]

It is worth recalling these appearances, not for complacency, but as a reminder that at the time it seemed that one could often see no chance of doing more than ease the patient's path and hope for nature's curative powers. Sir William Osler, whose scepticism of most of the remedies of his day helped to create an atmosphere of 'therapeutic nihilism', wrote:

> To accept a great group of maladies, against which we have never had and can scarcely hope to have curative measure, makes some men as sensitive as though we were ourselves responsible for their existence. Those very cases are 'rocks of offence' to many good fellows whose moral decline dates from the rash promise to cure. We work by wit, not witchcraft, and while these patients have our tenderest care, and we must do what is best for the

relief of our sufferings, we should not bring the art of medicine into disrepute by quack-like promises to heal, or wire-drawn attempts to cure in what old Burton calls 'continuate and inexorable maladies'.[27]

Each generation confronts problems that seem to it insoluble. Yet the recollection of diseases, once regarded as 'continuate and inexorable maladies', that have now disappeared teaches us to look in a different way at diseases which today seem equally inexorable. If, as Thomas Browne wrote about the memory of those that have died, 'oblivion blindly scattereth her poppy',[28] so too, Nature seems blindly to scatter her seed-corn of future discovery. As we look at the intractable problems of today—chronic neurological disease, schizophrenia, senile dementia, most forms of cancer, congenital deformity,[29] deafness, blindness, arthritis, and autoimmune diseases, and many forms of tropical and veterinary disease as well as emerging new infections[30]—it seems right to say, *not* that they are insoluble, but that it is our task to seek the knowledge which may provide the seeds of solution either in this or in a later generation.

Benefits to animals

Many of the drugs and procedures that have been of importance for man are of equal effectiveness in animals. The veterinary surgeon has the same antibiotics and antiseptics; the same hormones, tranquillizers, local anaesthetics, and general anaesthetics; the same surgical procedures and principles of resuscitation or other life support; and many of the same nutritional supplements. If one compares the medical and veterinary pharmacopoeias,[31] over half of those in the veterinary pharmacopoeia come from human medicine; the remainder focus on the specifically animal infections and parasites. The veterinary pharmacopoeia should be compulsory reading for those doubting the animal benefit from animal research! James Herriot's stories about the life of a veterinary surgeon provide many illustrations, some of them all the more vivid because he was practising while the therapeutic revolution was coming in.[32] This extends even to relatively rare conditions: in dogs, bladder stone due to a congenital cystinuria has been controlled by the human remedy, penicillamine; the identification of a type of haemophilia in man with that in dogs has advanced knowledge for both, and there are many other similar cases of similar inherited disorders.[33]

But there are two main areas specific to animals, where the work was done specifically to deal with animal disease: major epidemic diseases of various sorts, and infection with worms. The attempt to produce vaccines against animal disease began almost as early as that for humans, with Pasteur's work both on anthrax and on rabies. Fifty years ago Sir Leonard Rogers was able to calculate that by then over 100 million animals had been saved by inoculation against anthrax and rinderpest (cattle plague), and a similar number by swine-erysipelas inoculation in Germany, since the treatments were discovered.[14] Dog distemper, which used to kill animals by the hundreds of thousands, has also nearly been abolished. Other epidemic diseases due to organisms related to typhoid, dysentery, plague, cholera, and gas gangrene have been controlled. Virus diseases that can be checked by vaccines include Marek's disease in chickens and a feline leukaemia, both proving important also in understanding the relation of viruses to carcinogenesis. The Veterinary British Pharmacopoeia now contains over thirty vaccines, and new developments continue. A recent one is the discovery of a drug treatment for an African disease, 'East Coast fever', due to *Theileria parva* carried by ticks: this has been killing half a million cattle a year by a febrile disease affecting the lungs and ending with the animal drowning in its own froth. A more sophisticated advance is the combination of animal work and molecular biology to produce a new vaccine for foot-and-mouth disease.

The other main area of disease is responsible as much for chronic illness as for mortality: namely infection with nematodes (roundworm), cestodes (tapeworm) and trematodes (the flukes). The roundworm can both damage the alimentary tract and get into the lungs; the tapeworm causes wasting; and the flukes, with their extraordinary life cycle, can produce very varied damage. For almost all of them today some sort of treatment is possible; and the greatest practical problem is how to make a treatment that is equivalent to the deworming of a domestic pet available on a far larger scale.

The reader, after reviewing such evidence, even if he acknowledges the transformation in the life of livestock and domestic pets, may nevertheless feel that animals' interests are still not taken properly into account, and that the benefit is essentially for an unworthy cause, namely to make it more possible for man to use animals more readily for his own purposes—food, wool, skins, sport, or as pets. Yet whatever happens to the animal in the end, it must be worth while to make its life more comfortable; what we *use* animals for is a separate issue.

A similar argument could be advanced against the sulphonamides and penicillin. At one time epidemics of cerebrospinal fever (meningitis) used to occur in young soldiers clustered together in barracks, and could restrict the concentration of troops at some point at the front. This was one way in which the therapeutic revolution aided the world's armies, as typhoid vaccination had done earlier. Yet who would turn the clock back because it helped the soldier as well as the citizen, just because the soldier's job is to fight? It is the same argument that would have led to the withholding of anaesthesia for childbirth, because the suffering was considered at one time to be essential to maternal moral welfare. It could also be argued, along the same lines, that in order to prevent promiscuity, chemotherapy should not be used to treat venereal disease. In all these cases there is a suffering human being or animal: we may wish it would go away, but in the real world, where one sees the ravages of cattle plague, the suffering of meningitis or of prolonged difficult childbirth, or the end results of venereal disease, who can do other than seek to remove them?

Benefits in tropical medicine

Anyone dealing with the benefit from animal experiment, or from medical research as a whole, for tropical medicine, must do so circumspectly. A routine jibe is that all that medicine has done is to substitute death by starvation for death by disease. The short answer is that a relatively healthy population is more likely to find ways of improving its economy than a diseased one. A more serious comment would be that made by L. G. Goodwin in his two tables (Tables 4 and 5 below) listing respectively the six diseases earmarked for special attention by the World Health Organization, and the proposed uses for the sixty-one new drugs developed in one particular year (1975)—none having anything to do with tropical medicine.[34] That must dispel any complacency, and the reasons are not hard to find. While the strategic knowledge required for a medical advance may come from university, institute, or industry, the actual drug development under today's conditions can come only from industry. It is difficult, unpredictable, extremely expensive, and takes between five and seven years; the product may be pirated by those who have done nothing to develop it; and the people who need the drugs are too poor to pay for them. So the field is hardly viable financially, and all research proposals must pay their way, whether to a nationalized or to a private industry. The problem, however, is like that of the other 'orphan' drugs for diseases too rare to provide a sufficient market to support production, so that in

Table 4. Six diseases selected for special attention by the World Health Organization

Disease	Number of people infected (millions)
Malaria	200
Schistosomiasis	200
Filariasis	300
Trypanosomiasis	8
Leishmaniasis	2
Leprosy	11

effect a subsidy is required. Fortunately it is at last being taken seriously at the only level at which it can be dealt with—that of co-operative international governmental action.

Given that background, the success achieved in tropical medicine, may seem all the more notable. Some are very familiar to the traveller: vaccines for smallpox, yellow fever, typhoid and para-typhoid, and cholera (rather poor initially, but now improving). Considerable help has been provided by establishing the nature of the vectors (insect or animal) which transmit a disease; and this can allow control, even if treatment is inadequate. Drug treatment for a long time relied on traditional remedies, quinine for malaria and ipecacuanha (emetine) for amoebic dysentery. But the Second World War provided a great stimulus, and new antimalarials, together with DDT to control the insect vector of typhus,

Table 5. New drugs introduced in 1975 (in those countries chiefly concerned with drug development, especially USA and Western Europe)

Central nervous system	15	
Anti-infective	12	(none for tropical disease)
Cardiovascular	7	
Anti-inflammatory	6	
Cancer	6	
Hormones	4	
Respiration	3	
Gastro-intestinal	3	
Others (including diagnostic)	7	

sulphaguanidine for dysentery, and dapsone for leprosy, led the way. Further antimalarials followed, with drugs for filarial infections, for schistosomiasis, and for kala-azar, trypanosomiasis, and onchocerciasis. Their success, however, is more impressive if one looks at the past rather than at what remains to be done. Drug resistance is emerging; some of the agents are distinctly toxic; and a pervading difficulty, almost as great as that of discovering a new drug, is that of discovering how to get medical care delivered to those who need it. Indeed the latter applies, too, to all those diseases that the industrialized world shares with the developing world; for instance, there are the usual bacterial infections. Prevalence of rheumatic fever remains high in India, Africa, the Philippines, Central America, Indonesia, and the Middle East.[35] Measles-attack rates of nearly 50 per cent have been recorded in young children in The Gambia, although it was shown to be reduced to 5 per cent after vaccination.[36] Poliomyelitis still paralyses seven in every 1,000 Ghanaian children.[37] Much more could be done simply by implementing existing knowledge more thoroughly.

Nevertheless, the outlook for tropical medicine has, in principle, been revolutionized. The *possibility* of control of almost any of the diseases concerned is now quite clear, and even where resistance develops, our understanding of the cellular mechanisms involved is already pointing to ways of counter-attack. In all this, animal experiment, along with clinical medicine, epidemiology, zoology, chemistry and biochemistry, and engineering, have all played their part.

We should note, too, the extent to which man has used himself as an experimental subject in such work. Humans have volunteered to be infected with yellow fever, poliomyelitis, malaria, infectious hepatitis, as well as with less hazardous yet hardly more appetizing roundworms and threadworms. A remarkable modern instance is the work by Ralph Lainson and his colleagues on leishmaniasis in South America.[38] The task they set themselves was to get convincing evidence about the 'vectors'—in other words, how the disease was carried. This they did first by acting as human bait for sandflies (standing at night stripped to the waist, and then collecting the various species). They then found which species could carry the organism, and surveyed its natural occurrence. They learnt how to keep the insects and to make them carriers, and studied how the insects could acquire the organism from existing skin lesions. Finally they produced an infection with one of the infected sandfly species in a volunteer. Figure 11 shows a sketch of

Fig. 11. The ecology of leishmaniasis of the skin in Brazil

Sandflies (*Lutzomyia* species) which have not yet laid eggs (NV) ascend tree trunks to the canopy where they feed at night (V1, V2) on primary 'reservoirs' such as sloth and anteater (1). Opossums (2) and a third sandfly (V3) may sometimes be involved. Engorged flies migrate down trees to lay eggs (ovipost: 0). During the day they move back to the canopy for the next blood meal, when infected flies transmit the infecting organisms (*L. braziliensis guyanensis*). If disturbed at ground level during the daytime or early night-time, some flies will feed on the nearest mammal, and terrestrial species (2t) may be infected. Man is a common 'victim' and host (3) because of his excessive disturbance of the

the organism's life cycle. The work was combined with similar work using hamsters and other rodents, in whom leishmaniasis occurs naturally. It constitutes an extraordinary detective story in the maze of subtle adaptations between various species of organism, insect, and animal.

Limited experiment and unlimited benefit

We must remember, too, one consequence of man's accumulative capacity. His successors tomorrow build on what he discovers today. It is true that ideas, theories, and interpretations evolve: some of the science of today will seem as bizarre to our successors as the humours or the old phlogiston theory of combustion do to us. But that does not destroy the value of the knowledge won. While Einstein may have corrected Newton, yet for the vast mass of everyday practice, Newton's laws are sufficient. Our understanding of what happens when we vaccinate against a disease is sure to change; but the fact that we can, if we wish, abolish other diseases in the same way that we have abolished smallpox, does not have to be rediscovered, nor how to anaesthetize or kill organisms with antiseptics, nor how to repair damaged joints, relieve the asthmatic, or ease pain. Whatever suffering was entailed by such discoveries was limited: but the benefit to knowledge and to practice stretches far into the future for millions of humans and animals yet to be born.

The only useful suffering?

If we look at the world as it is, we can be overwhelmed by the suffering it reveals. But that is too one-sided a picture. Man's personal cruelty to man goes along with great generosity and kindness. War shows ruthlessness and brutality, but also great courage. Nature may be red in tooth and claw, and, as has been said, 'The end of every wild animal's life is a tragedy'; but there is also much evident beauty and enjoyment. Nevertheless, it is still true, as Lord Justice Moulton said, that 'The greater part of pain were better not to be.'[39] So much of it seems useless, even when one has allowed for need of warning of bodily damage, or for the necessary price that freedom of action in a dangerous world calls for. It is all the more strange, therefore, that the suffering which carries the hope of reducing future suffering seems to be so bitterly attacked. What *use* is there in the suffering as a cat plays with a bird, or a dog worries a sheep, or in any of nature's predatory activities? What *use* is the pain of a knee-capping in Northern Ireland, or shattered limbs on a battlefield, or of the mugged victim, or of the traffic accident? Each points forward to yet more suffering, as revenge works on and while the puzzles of how to resolve conflict or to moderate human action still remain unsolved. But if we turn instead to animal experiment and to the medical and veterinary practice built on it, it is a different picture, with some prospect, stretching into the future, of reducing human and animal misery. Without such experiment, without the deliberate search for knowledge, all we could look forward to is the long, random, incompetent trail of nature's own experiments on human and animal life.

The 'test of deletion'

The reader is asked now to turn back to Figure 1 (p. 44). The arguments against animal experiment have changed little in a hundred years, and if they are valid now, they would have been valid then. Let us suppose you accepted them in the past. Where will you place your bar, or your restriction, to animal experiment? Which of the advances mentioned would you be willing to have dispensed with or to have delayed? What accompanying ignorance of the human or animal body, now removed, would you have wished to perpetuate? At each date you consider, the benefits and knowledge still to come were all hidden, just as future benefits are hidden from us today. The only warranty for further work lay in the historical record of what had been already achieved, the knowledge already gained, and the openings to new discovery that had been created. The position is the same today: the only difference lies in the richness of the past record.

Summary

1. The 'benefit from knowledge' is assessed by imagining that no animal experiment had been done for 2,000 years, so that we lacked all knowledge so gained. The resulting ignorance both about our own bodies and those of animals would profoundly affect our outlook, to each other as well as to animals: it would remove much of the impulse to animal welfare, and, since knowledge about the inanimate world could largely have progressed unimpeded, it would produce a remarkable asymmetry in the pattern of our

understanding—primitive as regards the animate, highly advanced for the inanimate. Deliberate failure to win new knowledge is the same as deliberate perpetuation of ignorance.

2. In assessing 'benefit from use' flowing from animal experiment, care is needed to allow for other causes of advance, and for the lessons learnt about the placebo response and the fact that one cannot rely solely on doctors' or patients' approval of a remedy. The lack of reliable morbidity data restricts us largely to records of changing mortality. Cogent illustrations of advances for which animal work was essential in a variety of diseases are cited, including infectious disease, cardiovascular disease, leukaemia and cancer, animal diseases, and tropical medicine.

3. One consequence of man's capacity to accumulate knowledge is that limited experiment gives rise to unlimited benefit: smallpox vaccine does not have to be discovered again. This introduces something in the nature of a multiplication factor in assessing benefit.

4. Of all the forms of animal suffering that exist, only that incurred in animal experiment, and the medical and veterinary practice based on it, offers the prospect of *reducing* future human and animal misery.

5. A 'test of deletion' is proposed whereby the effects of abolishing or restricting animal experiment now is assessed by considering the effect of a similar abolition or restriction in the past.

6

THE PATTERN OF DISCOVERY

The human mind is often so awkward and ill-regulated in the career of invention that it is at first diffident, and then despises itself. For it appears at first incredible that any such discovery should be made, and when it has been made, it appears incredible that it should so long have escaped men's research.

Francis Bacon, *Novum Organum*, Aphorism CX

The last chapter reviewed a small part of the benefits that have flowed from research to which animal experiment was an essential contributor. We may well admire and be grateful for it; yet we may still wonder what assurance there is for future success. Bacon's aphorism quoted above is profoundly true. What *has* been achieved soon seems easy, and is taken for granted; and only the teacher or historian gets insights into how difficult these achievements were. If we look to the unknown future, how are we to know, for instance, that *these* research workers provided with *these* resources will make responsible and successful use of them? It is, in fact, a familiar question: we ask, too, how confident are we that this doctor or this solicitor will give us good advice, or that this architect will build us a good house, or that this Member of Parliament will represent our interests? There, too, we look at their past records. But the question has a sharper edge when it concerns discovery, the bringing to light of what is, at present, unknown; and it is nearer to asking whether this musician or poet or painter is worth supporting. We have to face the fact that, in principle, we cannot say what will be discovered, or when.

Yet history is more reassuring than this. It is not simply that one can quote examples of the apparently trivial becoming important; such as the work on the pigments in the butterfly's wing (the pterins) becoming central to certain areas of biochemistry such as the synthesis of nucleic acids, or the way that the exquisite sensitivity of the back muscles of a leech to nicotine provided the technical gateway to the theory of chemical transmission. It is rather that it is very hard to say of any piece of work that *no* practical benefit can be envisaged as resulting from it. Hearing me make this remark at a press conference once, a journalist challenged it, saying that he had done some graduate research on the biochemical uptake of an amino acid into the retinal pigment of the rabbit's eye, and that it seemed to him to have been of little scientific and no practical use. Yet the answer was obvious. Either as a result of vitamin A deficiency, or as a congenital defect, a condition known as 'night blindness' can occur, due to a defect in or absence of the retinal pigment. The metabolism of that amino acid, or some disorder of it, could easily become a valuable technique or clue in seeking to understand these conditions.

The discovery–benefit lag time

Fortunately, we can turn from such anecdotal instances to an interesting and comprehensive study, illustrating how it is possible to do 'research on research'.[1] This was an investigation by two Americans, Julius Comroe, a cardiovascular physiologist, and Robert Dripps, an anaesthetist, both of considerable distinction, into the knowledge that led up to ten major advances in the medicine and surgery of the heart, circulation, and lungs. Their method was first to collect opinions from nearly 100 specialists about which had been the ten most important advances in the field. Then, with the aid of over 160 consultants, they reviewed the body of knowledge that it had been necessary to acquire before the ten clinical advances could reach their present state. In the process they examined around 6,000 published articles, identifying about 3,400 specific scientific papers for study. Of these 663 were regarded as key ones, underpinning the bodies of knowledge required, which fell into 137 separate lines of scientific investigation. It is worth giving some of their final conclusions:

(a) The public, including physicians and many scientists, still equate one important discovery with the name of a single man, e.g. polio vaccine = Salk. However in every instance that we studied, previous work by scores or hundreds of competent scientists was essential to provide the basic knowledge for the widely known clinical advance, usually attributed to one man.
(b) Of the 663 key articles essential for 10 major clinical advances, 41.6% reported research done by scientists whose goal at that time was *unrelated* to the later clinical advance; 41.6% sought that knowledge for the sake of knowledge. Such unrelated research was often unexpected, unpredictable, and usually greatly accelerated advance in many fields.
(c) Analysed in another way, of the 663 key articles, 61.5% described

'basic' research, research that was performed to determine mechanisms by which living organisms (including man) function or by which drugs act; 20% reported descriptive clinical investigations without any experimental work on basic mechanisms; 16.5% were concerned with development of new apparatus, techniques, operations or procedures; 2% involved review and synthesis of earlier work.

(d) Of the key research, 67.4% was done in colleges and universities or their medical schools and associated hospitals.

(e) Although most key research was done in clinical or basic science departments of medical schools, important contributions came from all of the basic non-medical science disciplines in colleges and universities (biology, botany, chemistry, mathematics, physical chemistry, physics, plant physiology and zoology) and from agriculture, dentistry, engineering, photography, and veterinary medicine, as well as from a variety of industrial laboratories.

(f) Except in unusual instances (e.g. the clinical use of X-rays), some lag always occurred between an initial discovery and its effective clinical application. We analysed the length of 111 such lags: 8% amounted to 0.1–1 year; 18% were 1–10 years; 17% were 11–12 years; 39% were 21–50 years; only 18% required more than 50 years for application.

The study was a pioneering one, so far unique in its scale and care, and can of course be criticized. It is rather US-centred. The judgements were made essentially by scientists with little help from historians. Despite all precautions, arbitrariness could not be avoided. Different results could well be obtained in other fields. The calculation of percentages is probably misleading. Yet one need only review the sequence of scientific reports that they assembled to accept the general truth of their conclusions: that science is a co-operative enterprise; that 'basic' research is essential for practical advance, just as (although they were not investigating this) practical discoveries often have a 'basic' significance; and that the benefit of a discovery in one area of science often requires information from other areas of science before its potential can be developed or even envisaged.

It is worth quoting a few particular examples in order of increasing magnitude of lag time. The discovery of X-rays was reported by an academic physicist in December 1895. Already in the following year many investigators were using it for medical diagnosis. By hindsight, though, one might say that it spread *too* fast, as the later evidence of the damaging effect of over-exposure to X-rays became apparent.

A typical 'goal-oriented' discovery was halothane, developed as a general anaesthetic free from the explosion risk of ether, more potent than nitrous oxide, and safer than chloroform. Around fifty years of academic and practical work on men and animals preceded its chemical synthesis in 1951; it took another five years for its safety, its potency, and its special properties to be worked out in animals, and for its production and formulation for clinical use. Today it is one of the most commonly used anaesthetics, in man and animals.

It used to be thought that it would be impossible to operate surgically within the abdomen, on the heart, or on the brain. The first repair of a wound in the heart, that of a rabbit, was made in 1882. It took fifteen years to achieve the first success in man.

In 1922 a disease in cattle was identified: the main symptom was severe bleeding, and was associated with the eating of spoiled hay that contained sweet clover (Melilotus officinalis). The chemical substance (dicoumarin) in the clover that was responsible was identified. It was found to antagonize the action of a vitamin—vitamin K—required by the body for making certain clotting factors. As has happened so often, a poison, once understood, offers the chance of being exploited as a therapeutic agent: in this case the possibility now existed of a drug to control blood-clotting in human disease, for which there was at the time no drug that could be taken by mouth. When tested it proved reliably active by mouth in animals and man, and its duration of action and safety in other respects were mapped out. Then, nineteen years after the starting-point, it received its trial as the first anticoagulant active by mouth, one of a group of substances now widely used to control blood-clotting and thrombosis.

The first sulphonamides (of which 'M & B 693', used for treating Winston Churchill's pneumonia during the War, is one of the most famous) were discovered in 1935. It was noticed by an investigator in 1937 that sulphanilamide produced a loss of sodium from the body. The trail then took many interesting turns. The beginning—chemotherapy with careful clinical observation—was followed by basic work on an enzyme in the red blood cell; that enzyme was then found in stomach and kidney, and it was discovered that the sulphonamide could inhibit it; the mechanisms were studied, other factors controlling urinary secretion were recognized, and related chemical structures were tested; finally, twenty-nine years later, industry had developed an important series of safe, orally active diuretics, used for controlling excess water and salt in the body. Another important dimension was then added, as knowledge about the value of controlling raised blood-pressure developed, with the gradual recognition of the effectiveness of these diuretics for such control.

As an example of a very long lag, one may quote blood transfusion. It was first done in 1667, by an early Fellow of the Royal Society, Richard Lower, from one dog to another. The first *safe* transfusions date from the First World War, nearly 250 years later. What occupied the interval? It was soon found that transfusing animal blood into man was dangerous, and it was not until 1818 that James Blundell, having established that the transfusion must be from the same species, performed the first transfusion of human blood: it was in a dying patient, but produced only temporary improvement. His first successful transfusion, using blood given by his assistant was in a case of haemorrhage after childbirth. But this again proved unpredictably dangerous, and further advances had to wait until a pioneer of immunology, Landsteiner, opened the way by discovering the blood groups in 1900. Then the discovery of reliable methods of blood storage and preventing blood-clotting, the development of suitable equipment, and the organization (under the stimulus of war) of donor services and distribution, allowed transfusion as we know it to develop.

Lastly, one may recall the work by Sir Henry Dale, mentioned in Chapter 3 (pp. 22–3). That study of the contaminants of ergot improved our understanding of shock, which was important in World War I a few years later. It led to the establishment of the theory of chemical transmission in the 1930s; and was important for the development of antihistamines in the 1940s and 1970s, and of some major drugs for controlling blood-pressure in the 1960s—a whole family of discovery–benefit lag times.

As we look back over such cases in detail, each step clearly needed other advances, depending in their turn on yet others. We see, as the fundamental reason for such lags, that the different parts of science do not stand alone, and that not only is it impossible to foresee the pattern of growth in each area, but that the future interactions are also hidden.

Despite all this, the unpredictability must not be overstated. It is not true to say that the outcome of research cannot be foreseen at all, and therefore cannot ever be commissioned. A more accurate remark would be that the greater the originality of a piece of work, the less predictable it is. A great deal of research in fact yields results whose general character can be, if not predicted, at least named in advance among the possible outcomes. It is this, in fact, that provides the backbone of research, fundamental or applied: the gradual movement, guided by logic and imagination, from existing knowledge to the adjacent possibilities. The unexpected is the uncovenanted bonus for those ready to receive it. But the capacity to force one's way to a particular solution remains limited. My own best insight into this came from personnel research into diving and submarine physiology during and after the war. When the war started, a diver, or a crew seeking to escape from a sunken submarine, were known to face certain hazards, particularly drowning, poisoning by oxygen or carbon dioxide, and decompression sickness; but the safe limits were unknown. So work on human volunteers established adequately for the first time the pattern of oxygen poisoning, and the amounts that could be breathed for a given time without symptoms or convulsions. It also discovered the amount of carbon dioxide required to produce unconsciousness. The design of breathing sets and the efficiency of soda-lime canisters for absorbing carbon dioxide were improved. Survival at sea was improved by appropriate study. Schedules were improved for slow ascent by a diver after particular tasks to avoid decompression sickness. But at the end of all the goal-oriented research, which yielded a much safer diving practice, the great questions were still unanswered: how is it that oxygen is toxic, or that carbon dioxide anaesthetizes, or that gas dissolves in the body under pressure and then separates out so unpredictably after coming to the surface? As a result, practice remains empirical to this day, and the greater freedom that would result from being able to control toxicity or bubble formation still eludes us.

The great question of judgement, therefore, both for the larger issues of deployment of scientific resources, and for the particular issue of the use of animal experiment, lies in the balance between the easily foreseen and the creation of openings for the unforeseen.

The dangers of preoccupation with benefit

The argument has been presented so far as though interest in knowledge and interest in benefit were two fluids that could be mixed in any order and proportion. Yet there is a sense in which knowledge always has a priority. Both scientists and those that commission their work have particular hopes; and there is always the danger that those results seeming to fulfil those hopes will receive the greatest attention, or even be incorrectly selected as typical. The nature of scientific work ultimately provides a protection, for the scientist's search is for that which any other scientist can confirm, something independent of the observer and his own wishes. This is the fundamental importance of the confirmatory experiment, otherwise seeming rather a redundancy. But formal precise repetition of each others' work is fairly rare. The usual procedure is to test some possibility that would be the case if the original conception were true, and only if this fails, to go back for a re-examination. So the possibility of a scientist moving ahead on insufficient evidence, believing he has found something 'useful', is only too real. The only protection is to forget about 'useful', and to ask simply, 'Is this the case or not?' Reliable benefit can only be built on secure knowledge.

The issue has come up in an interesting way in discussions of the ethics of clinical trial. The traditional procedure has been to set 'limits of statistical significance' to the results of such trial; that is to say that knowing, for instance, how variable the course of some disease in a group of patients can be, one does not accept evidence of an improvement in rate of cure unless it is bigger than a certain magnitude.

For instance, suppose that with some disease it is known that on average for a large number of patients 50 per cent get better in ten days. If you took a random sample of 10 such patients, the actual number getting better by that time would only sometimes be exactly 5, but would vary between 3 and 7 in most cases, according to the chance of whether favourable or unfavourable cases happened to be included. To be more exact, you would expect for trials of groups of 10, cure rates of between 34 and 66 per cent in two-thirds of the cases, and only one time in twenty would the cure rate be outside the range 18–82 per cent. So if you were trying a new drug on such a group, you would regard a cure rate of 66 per cent as no more than encouraging; but a 90 per cent cure would be quite convincing. (Equally a cure rate of only 10 per cent would be convincing evidence that the drug was not merely useless but harmful). Normally a difference is required greater than that which would be expected to occur 1 in 20 times by chance. (Those odds are not unreasonably strict: to toss four heads in a row—which many people will have done in an idle moment—represents a chance of 1 in 16.) Yet it has been claimed that it is ethically wrong during a clinical trial, once *any* apparent advantage of one treatment over another (say 51 per cent cure against 50 per cent) has been shown, not to give to all patients thereafter the apparently better treatment. That particular case does not merit serious consideration; but as the apparent difference between the treatments increases, or as the odds of the treatments being equally effective move from near evens to 1 in 3 and upwards, although still short of the 1 in 20 level, the question whether the trial should be abandoned becomes sharper and more awkward. So, too, those interested in animal welfare might ask whether animals could be saved by reducing the reliability of the knowledge to be gained; in fact, in a later section on toxicity testing such an argument will be seen to have something to offer.

The heart of the issue is probably quite simple: it is a question of deciding whether the experience gained is to be used as the basis of future action or not, and if so in what way. If firm knowledge is not an objective, if a physician is willing to use a treatment which is still quite likely to be the inferior one, if the scientist is willing to go ahead with a grossly unreliable index of toxicity, that is one thing. But if something is to be of use to others, if some security in one's knowledge for the future is required, there is no choice but to obtain that security.

The whole animal as a superb detector

The reader may well accept the argument thus far: that discovery is hard to foresee, that science is really a family of sciences, that basic and applied research are both needed, that they interact in subtle and complex ways, and that one must not be impatient. You could still ask why it is that with so much knowledge won it is *still* necessary to do experiments on whole animals.

The easy, yet correct, answer is that living organisms are so complex that we are far from being able to solve our scientific problems on isolated parts. We shall return to this when we discuss alternatives (Chapter 8), but some explanation is desirable now. This complexity of living organisms arises in several ways (further knowledge may reveal yet more). First, the blood is a very complicated fluid. As well as carrying oxygen, it contains nutrients required for the various tissues (absorbed from the intestine, or already processed by one organ and being passed to another); hormones, whose amounts vary with the state of the body; controlling proteins of various kinds, binding or transporting various essential substances or controlling water flow to the tissues; and the electrolytes (salts), controlling the responsiveness of the tissues, their water content, and their acidity. (The blood could, for a gardener, be compared to a particularly satisfactory circulating potting compost!)

Secondly, various tissues (especially liver, lung, kidney, and the lining of the small intestine) engage in a variety of biochemical transformations affecting the body at large: this may be the making available for other tissues of sugar, particular forms of fat, or special proteins; or it may be a protective mechanism, taking some substance which is potentially harmful and converting it to a less harmful form ready for elimination.

Thirdly, perhaps the most characteristic feature of a living organism is its power of adaptation. This is 'vitally' necessary in a literal sense; for the organism's continuing life depends on its capacity to maintain the special features of its normal existence—its body temperature, the correct supply of electrolytes and nutrients, and effective elimination of waste products. Any perturbation immediately leads to responses calculated to restore normality—the so-called 'homeostasis', or in the words of Claude Bernard, the French physiologist who first drew attention to it, 'the preservation of the internal environment'. These responses are extraordinarily varied: they may take place in seconds, like the blanching of the skin to minimize heat loss in the cold; or go on for weeks like the change in blood formation after bleeding or exposure to the lowered oxygen supply of high altitudes. They may involve almost any tissue, any hormone, or any part of the nervous system. Finally, and especially in respect of the nervous system, there are profuse long-distance connections between the different parts of the body.

The result of all this is two-fold. The first is the obvious point that isolation of any organ or any bit of tissue robs it of participation in this integrated physiology; and our success in making experiments on such isolated portions depends entirely on our capacity to replace artificially the tissue's normal needs and environment. The history of tissue culture records the progressive recognition, still incomplete, of all the conditions and all the nutrient substances required in the fluid bathing the cells. Even today, foetal calf serum is often required for the unidentified factors it contains favouring growth in tissue culture.

But a second result is that it is the whole organism in particular that gives us our greatest chance of seeing the unexpected. It already contains all the mechanisms, known and unknown, that we seek to understand; it already has all the astonishingly sensitive recognition systems for the various known and unknown hormones and chemicals of the body; its adaptive responses represent an indicator system that something we may not have been aware of has changed. The more we move to work on isolated parts, the more we are restricted simply to those possibilities that we have thought of. It is because of this that the whole animal must be used for experimenting with a new drug, or exploring the physiological action of some newly discovered body constituent, or investigating the function of a newly identified gland or cluster of nerve cells in the brain.

Individuality and variability

The other side of the coin of the complexity of living organisms is that each one is an individual, and that they therefore vary. With so many influences at work, as well as normal genetic variation, it is inevitable that responses to an experimental test will be different from one individual to another. This is what makes biological work fascinating; but it exacts a price. Where a chemist can measure the melting-point of a compound once and for all, the biologist will find, for instance, that each animal has a slightly different body temperature from every other, and that it varies with time of day, activity, and so on. To reach any general conclusion he must therefore find ways of minimizing the variation and of allowing for it. It also makes confirmation by other investigators all the more important.

Biological and medical statisticians have in fact made major advances in statistical theory in tackling the problems created. The problems always arise in the same general form, involving three factors: the variability of the response being studied (which can sometimes be usefully broken down into identifiable causes), the accuracy required, and the number of observations to be made. The variability of different biological characteristics is, in itself, very interesting. Sometimes the variation is small, like the level of sodium in the blood which is regulated within 1–2 per cent. Others, like human height, vary, but only within 5–10 per cent—for instance, the British Association found that, for those born in 1883, two-thirds of adult males had heights between 64 and 71 inches. Other characteristics are very variable indeed, such as human hair length or male hairiness. The interest lies in the fact that if some characteristic is relatively constant, then there must be some biological mechanism regulating it rather tightly; so we have a direct indicator of the existence and effectiveness of the 'homeostasis' mentioned earlier, and we can see, equally, that hair length is variable because there is little reason to control it (short of special Army discipline!).

The extent of the variability of a response and the precision required by the investigator are thus of key significance for animal experiment, essentially dictating the number of animals that must be used. Sir Richard Doll has remarked, 'There was a saying among my colleagues when I began the first year of preclinical studies to the effect that if it moves, it's biology; if it changes colour, it's chemistry; but if it doesn't work, it's physics. Some years later a friend added that if it sends you to sleep, it's statistics.'[2] Any teacher will confirm how much more than a grain of truth there is in this, but this book is no place for developing the statistical ideas that have been evolved by biomedical statisticians. It is, however, worth mentioning three points. First, a very great deal of effort, over fifty years or more, has gone into the task of discovering how to get the maximum of reliable information from a minimum of observations, both by finding ways of controlling variability or allowing for it, and by using experimental designs with the best mathematical basis. Secondly, estimating the variability of a response, as well as being of some interest, is itself useful—it helps you to design later experiments better, and it can (for instance) warn a physician who is going to handle a new drug whether a small increase in dose is going to produce a big or only a small increase in effect. Thirdly, in the end there is no avoiding the fact that the precision of a conclusion and the number of observations made have to be traded off against each other: if you wish to be very precise, and to have confidence in that precision, biological variation is such that a good many experiments must be done. The point is probably already familiar to the reader from experience with opinion polls.

Summary

1. The argument in this chapter turns to our expectation of future benefit. We see, first, a sort of predictable unpredictability. While we could well expect that work on an anaesthetic or on transfusion technique or on surgery of the heart would lead to an obvious clinical application, we could not expect that a physicist would generate a standard medical diagnostic tool, or that a study of a disease in cattle would be the source of an important anticoagulant for man, or that the noticing of an effect of a drug for treatment of infection would lead to a diuretic and better blood-pressure

control. The greater the originality, the less the outcome or its timing can be foreseen.

2. We have to recognize the danger of being preoccupied with benefit rather than knowledge. Knowledge has a kind of inevitable priority, and our confidence in the expected benefit depends on the soundness of the knowledge.

3. Because the body is so complex and there is so much we do not know, work with the intact organism remains important; it is a superb detector of the unforeseen.

4. The very individuality of men and animals creates variability of response, and hence the need to make not one, but a number of observations. The precision required from a result inevitably has to be 'traded off' against numbers of animals used; and nobody seriously interested in animal welfare can afford to be unfamiliar with the statistical techniques developed by biomedical statisticians to make that balance the best possible.

7

PAIN, SUFFERING, AND LOSS OF ANIMAL LIFE

When considering the costs incurred by the advances in knowledge and practical benefit involving animals, the chief issue in the past was the infliction of pain, and we have already outlined some of its features (p. 18). Today concentration on pain alone is justly regarded as too narrow, and we need to take account of less tangible experiences: 'suffering', 'misery', and 'distress'. In addition, but also as a principle arising from the concept of a 'right to life', loss of life or disturbance of normal way of life may also be issues. With all these, great difficulties of definition arise, and Sir Thomas Lewis's dictum about pain is appropriate: that it 'cannot be defined, but is known by experience and illustrated by example.'[1]

A possible categorization of pain

Yet despite all the difficulties it seems worth attempting some degree of categorization—first of pain—if only to fix our ideas. First we can readily accept that everyday life for men and animals alike cannot be totally free of discomfort. Whether in exploration, or play, or with a posture held too long, or with minor disease or trauma, there is a certain level of discomfort which is unavoidable, which no one would propose to treat and which does not interfere with normal activity. These could be termed the 'everyday' discomforts. In human life we can recognize a next higher level, where normal activity can continue, possibly with some effort, and where to take an aspirin or an equivalent pain-killer can seem appropriate. Examples would be headache, menstrual pains, 'rheumatism', a pulled muscle, or a sprain. There is a characteristic response in animals that seems analogous, the rather misleadingly entitled 'writhing' test in mice. Here some mild irritant, that would not be felt on intact skin but would on an abrasion, is injected into the peritoneal cavity (the lining of the abdomen). The response is a posture as though the animal was trying to rub something away from its abdominal wall, with episodes of a characteristic movement and posture of its hind legs. But while it is doing this it can still be seen to be exploring any novelty in its environment, head turning and whiskers twitching, in an entirely normal way. The response is of interest because it was the first animal test that was found to be sensitive to aspirin. Perhaps, therefore, we can characterize such pains, which modify but do not seriously interfere with activity, and are treatable at the aspirin level, as 'minor'. The third category is 'severe' pain, where the pain is overriding and other than momentary so that activity tends to centre round it while it is experienced or liable to be experienced. It is for this that the opiates are needed; they are effective in animals as well as man. Examples in man would be a coronary thrombosis, a fractured limb, some cancer pains, some neurological pains, and some pains after surgical operation.

Suffering

If pain presents problems, suffering, misery, and distress are even less definite. The best we can do, it seems, is to extrapolate from human experience: post-influenzal misery, gastro-enteritis, and chronic infection form one category. Fear, anxiety, and frustration form another. Another term, which is widely used, is 'stress', but it has become almost meaningless unless an illustrative example is quoted. The term became more common when it was adopted by Hans Selye in his development of the concept of a 'chronic adaptive syndrome'. He assembled evidence that while 'stress' might have many different causes—such as injury, disease, or exposure to extremes of temperature—yet a common response could be characterized. It could begin with the general emergency reaction, involving activation of the sympathetic nervous system and release of adrenaline and noradrenaline; it would then go on to release of a hormone from the pituitary gland, ACTH, which in turn stimulates the adrenal gland to produce the shock-preventing 'corticoids' such as hydrocortisone. If this adaptive response was not sufficient, and the stress continued, a final stage of 'exhaustion' would be reached with impairment of many body faculties involved in growth, reproduction, resistance to disease, and general activity. One particular result of this approach was that it suggested that one might use the level of adrenaline or ACTH or hydrocortisone in a body fluid as objective measures of stress. This could be extended, on the assumption that the situations producing stress are those that produce suffering, to using these objective tests as a measure of suffering.

Assessment of pain and suffering

The problem of knowing how much pain and suffering an animal experiences, whether in experimental work, in nature, in animal husbandry or veterinary practice, culling wild life in game parks, or in domestication, seems at first wholly intractable. Marion Dawkins, however, in a recent book, *Animal Suffering*, has performed a most useful service in bringing together the sort of evidence that can be brought to bear; and it is worth summarizing some of her suggestions.[2]

First, we may consider how far the animal's life has been changed from its natural course and the implications of this. For the animal experimenter this is particularly applicable to the general care of animals bred and housed for experimental purposes, and for long-term experiment. It would be wrong to assume that natural life is ideal, and that any deviation from it is necessarily adverse. Presumably domestication of an animal must be regarded as an unnatural state; yet to release a domestic animal into the natural life in the wild could well lead to its rapid death at the hands of predators. The study of 'imprinting' has shown how critical are conditions in very early life, so that the conditions of its breeding may be the best pointers to its needs in later life. While animal freedom is evidently desirable in general, it is not always clear that animals necessarily suffer if they cannot behave as they would in the wild. Indeed human protection could make them better off, for natural life cannot be regarded as free from suffering: witness the high mortality, for instance, of a songbird in the wild (one to two years), compared with that in captivity (eleven years or so).

Second, we may look at the general health of the animal: its eating and drinking, growth, physical appearance, and reproductive performance. Of course, a particular index such as favourable growth is not, in itself, decisive; putting on weight might be simply an index of excessive confinement. Similarly, lack of any overt effect on health does not exclude suffering; if the suffering is brief, there may be no time for evident health defects to appear.

Third, it may be possible to establish physiological measures that can be taken as signs of suffering. But these are not in themselves decisive. There are considerable uncertainties in correlating actual levels of adrenaline or adrenal corticoid levels with degrees of suffering, both because of the variability between individuals, and because of all the other higher influences that affect suffering (such as memory, expectation, distraction). In addition, release of these hormones can be regarded as part of the general 'homeostatic' regulatory mechanism of the body, and their activation does not always reach consciousness.

Fourth, we may look at the animal's behaviour, not only obvious avoidance or reflex responses, but other changes, such as change in the grooming of fur, or signs of conflict, apathy, unnatural aggression, or abnormal biting or other stereotyped pattern.

Fifth, it may be possible to determine an animal's own preferences by the choices it makes. This connects with the techniques used extensively in experimental animal psychology, whereby animals are trained, either by an aversive stimulus or by some reward (such as food or drink) to perform certain actions. If one assumes that what it chooses is in fact preferred, and that what it avoids would involve suffering in some sense, then one is getting direct information. Experiments of this sort showed for instance that, contrary to human expectation, hens prefer a fine mesh flooring to a stronger, wider gauge metal flooring. There are, of course, difficulties. It would be very hard for man to anticipate all the factors influencing animal choice. It is also far from clear that animals would choose what was for their own benefit (the sight of a cat trying to remove the dressing over an infected eyelid is sufficient example). Yet there remains an area where useful evidence could be obtained.

Finally, we can extrapolate from human experience, but the dangers of doing this are obvious. Medawar cites the charming example of the little girl who thought a frog in the garden needed warming up, because it was cold.[3] Yet, if combined with sufficient appreciation of the animal's normal physiology and life-style, particularly when we are considering other mammals, such arguments from analogy seem legitimate.

There are, therefore, various approaches, all potentially informative though none decisive in themselves. It falls to those concerned with animal welfare to piece this evidence together. In doing so, experience is at a premium. It requires training even to know how to pick up an animal without causing it distress. It is a field in which the veterinary profession can give excellent advice.

The development of skill in analgesia

We have possessed efficient analgesics ('pain-killers') for many years. The prototype of the powerful analgesic is morphine, the main active ingredient in opium, which has been known for centuries. Of the milder analgesics the prototype is aspirin, discovered in 1899. But both drugs have other actions which complicate their use in relieving either human or animal pain. Morphine, as well as being liable to produce craving, tolerance, and withdrawal symptoms if given repeatedly, also constipates through its action on the bowel's movements, depresses the breathing, affects the circulation, which can cause fainting, and may produce a reaction akin to allergy by direct release of histamine in the body. Aspirin is commonly an irritant in the stomach, and in large continued doses can damage the kidneys, as well as producing allergic reactions in some subjects. Pharmacological research and industrial development have done much to widen our choice of drug, producing alternatives that are active by mouth or injection and have varying potencies and addiction liability, varying duration of action, and various degrees of associated action. The hospice movement, the development of pain clinics, and work by anaesthetists have all stimulated research into the control of pain, and many useful ways of combining analgesics with tranquillizers or other drugs have been found. It is probably now true to say that, in principle, no one need suffer prolonged severe pain, even if for certain conditions some discomfort is bound to remain. The essential problem in human practice is how to make available the knowledge and skill required on a wide enough scale.

If we now turn to pain in animal experiment, it is clear that *in principle* a good deal could be done to mitigate it. But two principal difficulties must be recognized. First, much of our detailed knowledge of pain control is limited to human experience; and it by no means follows that the details of analgesic treatment will be identical in animals, or even the same from one animal to another. In particular the duration of action of a given dose may be considerably different. Secondly, the associated actions of the analgesic could well distort the results of the experiment. In turn this could well increase the number of experiments done, because extra experiments would be required to disentangle the effects of the analgesic from those of the test procedure.

The RSPCA has in recent years been reviewing the scientific literature for experiments in which pain and suffering are evidently involved. Other bodies have made similar studies from time to time. In general, severe pain is rarely, if ever, produced. But we can tentatively identify some areas where it seems likely that the use of suitable drugs offers scope for reducing pain and suffering: in post-operative care; in research on infections and their treatment; in cancer research where tumour growth is substantial; and in tests where chronic exposure to a drug is necessary and where the drug produces some adverse effect.

It would be useless to insist, at once, that in such instances analgesics must now be used. Adequate knowledge of how to allow for the presence of the analgesic does not exist, so that many more experiments would be required; and it is not at all clear whether there is yet sufficient knowledge of potency and duration of action of suitable drugs for their safe use in all the species involved. In recent years, the concept of an 'institute' for the development of alternative methods has been forcefully advanced. The reader may feel, however, after reading the account of their development and scale of use (pp. 102–7), that this would hardly be a useful investment. More useful, it seems to me, would be the deliberate funding, at suitable centres, of research into methods of relief of pain and suffering in animals, particularly in the context of experimental work. To some extent it would be rather severely *ad hoc*: to discover, for instance, the most specific analgesic, with fewest side-effects, lasting the right length of time for the relief of post-operative discomfort in rat, rabbit, cat, or dog; or a drug which relieves any suffering that arises in cancer research without distorting the process of tumour growth or of tumour destruction by some test drug. Systematic knowledge of potency and duration of action for a range of drugs in a range of species would itself be of great use, particularly if associated side-effects could be delineated. Similar considerations apply to the use of local anaesthetics which can be used when a source of pain is strictly localized to the area supplied by some identifiable sensory nerve. Such work would, however, be likely also to yield some more general benefit. There has been considerable improvement in our understanding of the mechanisms of the production, recognition, and control of pain; and it is likely that the experiments could be done in such a way as to exploit and further advance that understanding. Further, one must remember that, while most experiments are done on normal healthy animals or men, in fact both humans and animals when they receive drug treatment are often diseased or receiving other drugs. To test an anti-cancer drug on an animal receiving an analgesic may well be a more realistic model of treatment in man, even if it also becomes more complex.

It is with work of this sort in mind that the idea of categorizing pain was outlined earlier. It would be foolish to use a sledge-hammer to crack a nut, and one of the tasks would be to develop pain-relieving regimes appropriate to the nature of the suffering involved.

As with the assessment of pain, this is a field where veterinary expertise could well be especially helpful. Realistic appraisal of suffering is obligatory; and without adequate experience, a layman might, for instance, propose post-operative analgesic drug treatment for a rat in whom a recording device had been implanted under anaesthesia, which might impair the return of eating, drinking, and normal activity, and actually prolong rather than mitigate post-operative distress.

Summary

1. Since it is necessary to assess in some degree the pain that may be incurred both by animal experiment and by the failure to do animal experiment, a possible categorization of pain is proposed: (a) 'everyday' pains, that are accepted as not worth treating and not interfering with normal activity; (b) 'minor' pains, that may modify but do not dominate activity, responsive to aspirin-like drugs; (c) 'severe' pains, where pain is overriding, calling for opiate drugs. It is agreed that categorization offers great difficulties; yet evidently some pains are worse than others, and 'bench-marks' could be helpful. There seems no way at present to categorize suffering or distress.

2. While pain and suffering in another being is, strictly speaking, unknowable, yet for practical purposes there are a number of lines of evidence of their presence and intensity which deserve attention: (a) evidence of normal pattern of life; (b) impairment of general health; (c) physiological indices of 'stress'; (d) changes in behaviour patterns; (e) expression of an animal's own preference; (f) extrapolation from human experience. No single index is reliable; but attention to the various aspects, combined with experience with animals, can improve the assessment.

3. It is suggested that more research could usefully be done to discover suitable procedures for minimizing any pain and suffering involved in experimental work. The two main difficulties are that present knowledge of analgesic techniques is largely based on human trial; and that the introduction of analgesics into an experimental procedure could, by complicating it, increase the number of experiments required. Deliberate research for the

development of analgesic procedures of the highest specificity, and suitable for different species, could reduce these difficulties.

4. Veterinary expertise should be deployed to assist in the further development of skills in anaesthesia.

8

THE ALTERNATIVES TO ANIMAL EXPERIMENT

It may be accepted that animal experiment has been justified in the past, and will be in the future, yet the concern remains that *unnecessary* experiments should not be done. We must turn, therefore, to ways in which this might be avoided.

Avoidance of repetition

One charge is that the same animal experiments are frequently repeated, for no good reason, and that considerable savings could be made by stopping this. There are some obvious points to make. First, some repetition is absolutely necessary: without it a result can easily be dismissed as 'unconfirmed'. It is also essential, if others are to build on previous work, that they establish a genuine link with it, which involves, at the start at least, similar experiments. Often, too, control experiments of some sort are needed, to be done under the same conditions as some test procedure, and these may well appear repetitive. One must recognize, also, the forces working against such unnecessary repetition: cost; waste of time; and the existence of many people, such as heads of departments, other scientists, and grant-giving bodies, only too willing to point these out. But the question can still be asked, whether such redundancy occurs in practice or not. A short answer is that the Littlewood Committee (1965), in its very thorough study for the Home Office of British practice in experiments on animals, considered this point specifically, took evidence about it, and concluded that the risk of unnecessary repetition of experiments was small and the scale of duplication not serious.[1] Their discussion of the whole matter, about which evidence was widely taken, is well worth reading.

Since then, the literature has of course expanded very greatly, so the position might have deteriorated. But computer aids have now also become far more widely available and much more powerful. Should it not now be possible with such aids to avoid any repetition at all? The question of the efficiency of information retrieval was the subject of a valuable study in which 14 bodies (2 university, 1 governmental and 11 industry) took part.[2] It concerned particularly the recovery of information in chemical toxicology and involved framing eight test queries, designed to represent the sort of information likely to be sought, sometimes very specific, sometimes rather general. The various participants then conducted a search using the sources available that seemed most appropriate, and using methods that would be used in real life (such as following up references and browsing over a wide variety of handbooks and textbooks; printed indexes; on-line data bases and data banks; and in-house material). The results and techniques used were then compared, and a consolidated list of references obtained, from which the value of various sources and methods of retrieval could be estimated. Some difficulties very familiar to the scientist were highlighted.

First, with one query most of the literature pre-dated the modern abstracts and data banks and was found only in old textbooks or individual papers. This difficulty will always be present to some degree, even as the methods of data storage and retrieval steadily improve; it will never be possible to bring *all* past information into the format used in the latest data-storage system.

Second, the required information could exist as a minor item, not indexed, as part of a large paper on some other topic.

Third, the 'on-line' type of retrieval, where a large list of brief titles can be obtained, was sometimes effective; but it sometimes failed to allow a desirable *evaluation* of the reference in the way that personal search may do, and could lead to a great excess of irrelevant material.

Fourth, there is a major dilemma in obtaining a balance between asking too general a question (leading to flooding with unwanted material) and making it too detailed (when specific details are liable not to be indexed, or not indexed in that precise form).

Finally, following on the above, it became apparent that literature search, as well as the selection of the specific material to be indexed, have both become highly specialized tasks. If this specialization develops too far, however, it would put effective search beyond the reach of all save major centres.

At the same time it seemed clear that there were two or three rather effective and commonly used principal sources of information which, taken in conjunction, could produce fairly readily up to two-thirds of what appeared to be the total literature available. This is rather reassuring. For if the question is 'Has the experiment

I propose to do been done before?' the chance of this is in any case small; and the information successfully retrieved is very likely to settle the major questions, namely whether there has been *any* prior study and, if so, the general properties of the substance in question. It needs to be realized that such searches will be conducted *in the light of knowledge at the time of the search,* which will be different from the earlier state of knowledge. This is the reason why precise reduplication of experiments is in fact uncommon, since the questions to be asked change as knowledge increases.

Use of the non-animal alternative

Historically, the interest in non-animal methods seems to have arisen in response to the experimenter's claim that physiological discoveries could not have been made by any other means. The challenge was met by suggesting other ways in which the discoveries could have been made. The alternatives suggested, for a long time, were chiefly 'mortisection' (in other words, dissection of dead humans or animals), clinical observation, and a priori thought—the methods used over past centuries, whose success but also whose failures are obvious enough. But the advance of biomedical science steadily made more analytic procedures possible, using chemical, biochemical, tissue culture, or other techniques; and their evident contribution to knowledge in turn suggested that yet more could be expected of them.

The idea of 'alternative methods', as a distinct entity, crystallized in the 1960s, and came to public prominence chiefly, perhaps, with the Bill brought forward in the House of Commons by (the then) Mr Douglas Houghton in November 1972. This sought to amend the 1876 Act by making it an offence to do an experiment on animals which could be done by alternative means not involving an experiment on a living animal. Although there was widespread agreement that alternative methods should be encouraged, the Bill was talked out at its third reading. One of the particular points made in debate was that it was sometimes very uncertain whether a method *was* alternative or not. A striking case was cited where a drug firm wished to use a chemical method for standardizing a pituitary hormone used in medicine; after two years of debate, however, the British Pharmacopoeia Commission insisted that biological tests should still be used, because of certain chemical uncertainties. Also, despite wishing to encourage such methods, it seemed unreasonable to make it a penal offence for an individual investigator not to have discovered, in the whole vast literature of medical science, that an alternative method existed. Finally, some measurements, for instance of blood adrenaline levels, could indeed be made by chemical methods, but only at considerable expense compared to tests on an anaesthetized rat. The effect of the new legislation would have been to penalize severely the smaller centres of work.

Since that time, there has been considerable discussion and a number of substantial meetings and publications on the subject, which have clarified the position. An excellent book by the late Professor D. H. Smyth gives a very full discussion of the issues involved, and much of the evidence available, after very wide consultation of all interested parties.[3]

For the research worker himself, it seems strange that he should find himself accused of neglecting alternative methods, or indeed that they should be regarded as something new to take account of; for it is the animal experimenters who have themselves been developing them and using them for decades. Rather than repeat the accounts of the limitations of such methods (see pp. 87–9) it is worth explaining their development and use.

There is a natural path that continually opens up before the experimenter, as opportunities for deeper analysis on simpler systems become possible as a result of knowledge gained in animal work. For instance, with a drug like digitalis the first experiments were on whole animals, since it was not known on what part of the body (brain, heart, blood vessels, alimentary tract, or kidney) it worked. Once the site of action is identified—such as the heart with digitalis—then work on an isolated, perfused organ becomes possible. After that may come the use of isolated tissues, then single cells, then subcellular constituents. The development depends on the previous animal work. The conclusions drawn from the more analytic work then also need to be checked in the whole animal to establish that what happens to pieces of cells in a test tube also happens in the intact tissue. All the alternatives I know of have traced this path.

Tables 6 and 7 list those methods that can be called in some sense 'alternative' that are familiar to me from the literature and from my own work. The word is used in various ways to mean either avoiding experiments on a living animal (although animals will have to be killed to obtain the material) or avoiding both live experiment and the killing of any animal (Table 6), or, at the

Table 6. Methods of experiment not involving whole animals

Isolated perfused organs using:	Isolated tissue or tissue sample using:	Isolated single cell using:	Subcellular constituents using:
Liver	Striated muscle	Fat cells	Nuclei
Muscle	Electroplax	Liver cells	Mitochondria
Heart	Iris	Neurones	Microsomes
Lung	Trachea	Glia	Lysosomes
Adrenal	Bronchi	Muscle	Synaptosomes
Intestine	Lung	Smooth muscle	Cell membranes
Skin	Intestine	Red cells	Actin/myosin of muscle
Spleen	Spleen	Leucocytes	
Kidney	Uterus	Platelets	
	Seminal vesicle	Mast cells	
	Vas deferens		
	Bladder		
	Salivary gland		
	Fat pads		
	Liver slice		

extreme, avoiding the use of living matter altogether (Table 7, p. 106).

Experiments with living isolated nerve and muscle began with Galvani in 1791, and continued through the nineteenth century. The first attempts at studying the heart in isolation go back at least to 1846. But success depended on knowledge of the essential constituents of the supporting fluid. This came only in 1880, when a lazy laboratory assistant, using tap water instead of distilled water in preparing solutions for use with a frog heart, allowed Sydney Ringer to recognize the need for calcium ions. That was the beginning of a trail in which control of acidity, glucose, other salts, and many other factors were gradually recognized, leading up to the tissue culture medium of today. Early attempts, dating from around 1900, were also made to perfuse isolated organs with blood, but could initially be attempted, for lack of any non-toxic anticoagulant, only by using blood that was allowed to clot but with the clot removed as it formed; an organ could then be perfused with this 'defibrinated' blood, which retained, of course, many of its normal properties. This was only partly satisfactory, since the process of clotting released pharmacologically active substances which could disturb the blood flow. Yet this, too, was a fruitful observation; for the attempt to characterize one of these substances led ultimately to the identification of 5-hydroxytryptamine

(5-HT), contained in the platelets of the blood; and 5-HT in turn was later revealed as one of the amines of profound importance in brain function. (A further by-product is work using the blood platelet, whose diameter is less than one-twenty-thousandth of an inch and of which we possess about a million million, as a 'model' and 'alternative' for the brain!) Another fascinating experiment found that to pass the defibrinated blood through isolated lungs 'cleaned it up'—an early indication of what is now recognized as an extensive metabolic capacity of the lung. Subsequently, reliable and safe anticoagulants—some occurring naturally in the body, like heparin, others made artificially—were discovered, and now play their part in the heart–lung and kidney machines of modern surgical practice which themselves evolved from the early perfusion experiments. As a result of these developments, most organs and glands in the body have now been studied in isolation either (if thin enough to allow sufficient oxygen to reach the depths of the tissue) by suspension in a suitable fluid, or by perfusion. These experiments do not avoid loss of animal life, but do avoid experiment on the whole living animal. Some preparations, such as the intestinal strip, have been remarkably useful because they possess two nerve networks (one controlling propulsive activity, the other probably secretion) as well as muscle and a wide range of pharmacological receptors; and a dozen preparations can be obtained from a single intestinal length, allowing considerable economy of animal life.

The next analytic step was tissue culture, inaugurated by Harrison's classical work culturing nerve cells in 1903. As with isolated organs and tissues, success has grown with knowledge of the essential constituents; and tissue culture is now a routine technique. In a parallel development, special tools were designed for manipulating tissues under the microscope, methods of electrical recording of great speed and sensitivity were discovered, and other physical, chemical, or biochemical techniques were developed so that experiments could be made directly on single cells a fraction of a millimetre in diameter.

The same analytic drive led to what is called 'subcellular fractionation' when, around the 1950s, ways were found whereby cells themselves could be gently broken, and their various 'organelles' within separated from each other. These included the 'mitochondria', little factories producing energy from our food in a form the cell can use; 'lysosomes', which degrade unwanted materials; 'microsomes', which convert foreign substances to less harmful and more readily eliminated forms; 'synaptosomes', containing the junction between two nerve cells, with the chemicals and receptor apparatus together; and cell membranes themselves, with all their machinery for responding to chemical and other stimuli, for transporting materials in and out, for secretion, and for recognizing foreign substances and other cells. From muscles the proteins whose relative movement results in the shortening of muscles were isolated (including actin and myosin), so that simple *in vitro* systems could be studied. Today, inspection of the literature shows that experimental research conducted on such isolated organs, tissues, cells, or cell constituents has grown steadily over recent decades, and now there are more experiments of this sort than there are experiments on whole animals.

The account thus far of the non-animal alternatives points to their successful aspects. But the earlier discussion (pp. 87–9) of the complexity of the whole body, and the continuing necessity for animal experiment must still be borne in mind. Two illustrative examples of the dangers of experiment on other than the whole animal are worth citing here.[4] The first concerns the discovery of the sulphonamides. P. Domagk's original experiment used a red dye, Prontosil Rubrum, and he tested it against a streptococcal infection in mice. It was subsequently found that it was inactive when tested *in vitro*, and that Prontosil owed its activity to being broken down in the body to produce the active material, sulphanilamide. It is singularly fortunate that Domagk turned first to trial in an animal. That was in 1935. Nearly fifty years later the same happened with an important new antiparasitic agent, active against roundworms (nematodes); this proved inactive by *in vitro* tests, but given by mouth to parasitized mice had a quite astonishing potency, being active with a dose of less than one millionth of a gram.[5] Perhaps the essential point is that it is wrong to view

Table 7. Procedures not requiring living matter

Structure–activity relationships	Analogue or computer modelling
Curare (1868)	Cardiovascular function
Nitrites (before 1900)	Anaesthetic uptake
Anaesthetics (1899)	Decompression sickness
Sympathetic amines (1906)	Drug metabolism and distribution
Barbiturates (before 1930)	Receptor function
Sulphonamides (1940)	Neuromuscular control
and many others	Nerve action potential
	Control of respiration

animal and non-animal methods as opposed to or competing with each other; in fact they serve different but complementary purposes.

Prediction of biological action from chemical structure: structure–activity relationships

Two other approaches also have a long history. Today's use of modern chemistry (based on quantum mechanics) to predict pharmacological action reaches back to 1868, when the first great generalization in the relation of chemical structure to action in the body was made by two Edinburgh scientists, Alexander Crum Brown and T. R. Fraser. They found that a particular pattern of combination of carbon and nitrogen atoms ('quaternary nitrogen') led to paralysis of the voluntary muscles similar to that produced by curare. A second great step was in 1900, when another generalization was made, that any substance liable to dissolve in fats to a given extent would be an anaesthetic. These two generalizations still guide us today, and were followed by a growing stream of similar generalizations as chemistry and pharmacology advanced. The great virtue of modern quantum chemistry is to give a far more realistic picture of how a chemical substance 'looks' to its receptive site than is provided by a chemical formula drawn on paper; and while the calculations are still imperfect, and are so laborious that they are only practicable with computer aids, they already provide in certain fields a useful predictive capacity. The strategic difficulty remains, of course, that their power is only as good as the biological information on which they are built.

Modelling

The modelling of biological phenomena is another area of interest. This too has a long history. The practical class offered to Cambridge medical students in 1876 included experiments with rubber tubes, glass connections, a pressure-recording device, and a hand pump, to teach the principles of the circulation.[6] That has been succeeded by later more sophisticated models, using analogue or digital computers to illustrate, say, the principles governing the rise and fall of the amount of anaesthetic in blood and brain if ether, chloroform, or nitrous oxide (laughing gas) are inhaled; or how the action potential arises and travels along a nerve or muscle when they are excited into activity; or how the centre in the brain controlling our breathing may switch alternately between breathing in and breathing out; or the mechanism by which our eyes or hands track a moving object. But some caution is needed in stressing the use of such models for teaching. Anyone who has done it (or examined students subsequently) comes to realize that unless the model is very simple and vivid, too much time necessarily gets spent on explaining the analogy and justifying its validity, at the expense of time devoted to the actual biological principles to be demonstrated. Such models, too, convey a rigid mechanical impression of biological responsiveness, a far cry from the individual and infinitely variable pattern of real life. The importance of such modelling can, however, be substantial for research purposes—essentially for exploring the results of some theory over a range of variables. But here, too, the models are only as good as the biological data fed into them. As is so true of statistics also, 'Garbage in, garbage out'.

Are alternatives sufficiently used?

Despite the history of the use of these methods it may still be thought, however, that there is insufficient pressure on the animal experimenter to use them. It is impossible to meet such a comment other than by pointing to the pressures already in existence, and to evidence that they are in fact used. The pressures for their use are: their scientific usefulness; their relative cheapness (usually, not always); the ability to generate more standardized testing procedures (of especial importance in industry); the formal requirement by the Home Office and bodies such as the Medical Research Council to verify that an animal experiment cannot be replaced by one avoiding animal use; and, not to be forgotten, the feelings of humanity, which the animal experimenter has no less than other people. Objective evidence for the progressive use of such methods as they have become available can be found in the Home Office Annual Return, which gives the number of experiments and the number of licensees.[7] From this it can be calculated that over the fifteen years between 1968 and 1982 the number of animal experiments per licensee has fallen by 50 per cent from 405 to 203. Yet the output of biomedical research by any test has greatly increased.

A second interesting calculation is a comparison of animal use with research expenditure, shown in Table 8, covering the last twenty years. One may take as an indicator of general medical

Table 8. Comparison of animal use with research expenditure

	1962–3	1964–5	1967–8	1971–2	1978–9	1981–2
MRC budget (£ million)	6.27	9.63	15.1	25.1	61.8	106.5
Retail price index (Jan. 1974 = 100)	53.5	57.1	63.4	82.9	207.2	277.3
MRC budget (Jan. 1974 £s)	12	16	24	30	30	38
Animals used in UK (millions)	4.2	4.8	5.2	5.3	4.7	4.2
Animals used nationally per MRC 1974 £	0.35	0.30	0.22	0.18	0.16	0.11

research the expenditure by the British Medical Research Council, whose support extends across the whole spectrum from the molecular to the clinical.[8] That expenditure needs correcting for inflation; and the third line of the Table gives the MRC budget in 'real' terms, namely 'January 1974 £s'. The fourth line gives the number of animals used over the period in Great Britain. The last line then takes this expenditure as an index of national biomedical research activity, which is, if anything, conservative in relation to industrial expenditure, and shows the animals used nationally per 'MRC 1974 £'. It will be seen that it has declined over the twenty years by about two-thirds.

Among other ideas, there has been considerable pressure for the establishing of some institute dedicated to the development of alternative methods, if necessary by diversion of the requisite funds. It is very doubtful that this is really needed, when one thinks of the proliferation of laboratories dedicated to cell biology. It would be more sensible to let such methods evolve from the scientific context in which they are needed, rather than to try to develop them *in vacuo*, with no answer to the question 'Alternative to *what*?' But there is one strictly practical point, which arises very clearly in questions of standardization of drugs by biological test. With these, satisfactory biological methods have been worked out, and doctors or vets and regulatory authorities alike rely on the products tested in this way. If some non-animal method is introduced, it can be accepted only if it is at least as reliable as the previous method; and to prove that, an extensive comparison between them is necessary, involving animal experiment that would otherwise not have been done.[9] Care is therefore needed, for if the new procedure proves inadequate all the extra animal experiment is wasted. In fact, both for humane and other reasons, there is considerable continuing interest in such moves, in both industry and research institutes. It is doubtful if, on the research side, much more could be done without incurring considerable additional animal use. Of much more interest, however, as is discussed later, is how far regulatory bodies can relax over-stringent standards, reducing animal use in that way.

The pattern of growth of new remedies

Such findings may be regarded as rather encouraging. Yet those who are concerned that we do not lose or delay future medical advance may also wish to be assured that the advance of knowledge is not being hindered. It is not easy to assess this, but one approach is simply to review the rate of introduction of new remedies. This is, indeed, worth a digression, since it may help to place modern therapeutics in a wider context. If we review the old pharmacopoeias we can recognize a variety of stages.[10] As we can see from Figure 12a, the first pharmacopoeia in this country, prepared by the Royal College of Physicians in 1618, contained over 2,000 remedies. The initial stage was the pruning of the traditional herbal and other remedies—most of them both ineffective and innocuous, and some quite bizarre with scores of herbal and other constituents. By 1746 there were only about 650 items, not only fewer but much less complex. This history really represents a sieving out of man's experiments over the centuries with the herbal and other substances in his environment. The uncritical appeal to such remedies today often ignores those long centuries of unrecorded trial and error, by which the useless or harmful were shed, but which effective drugs like opium, digitalis, or belladonna easily survived.

The next stage may be taken as the first step towards obtaining the pure active principles, free of inactive or harmful dross. So we see (Figure 12b) the rise of the tincture and the extract, each providing a way to separate out the active principles according to physical properties—fat-solubility or water-solubility respectively. This phase lasted a long time, even up to the 1930s. During this period the Pharmacopoeia of the Royal College of Physicians merged into the British Pharmacopoeia as we know it today, with its first edition in 1864.

Knowledge of microbial infection and of methods of disinfection, together with the introduction of the hypodermic syringe, made hypodermic injection of a few substances possible in the latter part of the nineteenth century; but injection did not 'take off' until chemical and pharmacological advance allowed the use of pure active substances in the 1930s. This was catalysed by Henry Wellcome's insight into the virtues of a convenient form of medication of constant stable composition. The word 'tabloid', which he patented, introduced a new word into the English language. The era of pure substance, in injection and tablet form, began. The age of the tincture and the extract entered its last stages.

Two new factors now appear (Figure 13). The first, from 1932, is the use of biological tests, which were used to standardize those substances which still baffled the skill of the chemist. It was thought in the 1950s that the number of these might be levelling off. But in fact the rate of discovery of biologically active substances still runs ahead of our knowledge of exact chemical constitution: knowledge of 'activity' can still run ahead of knowledge of 'structure'. The other factor is the growth in application of synthetic chemistry. We have to be a little arbitrary about our definition of 'synthetic'; and the line drawn in the Figure to represent 'unnatural' drugs counts those not obtainable *as such* from natural sources (beginning with ether and chloroform). Most are wholly synthetic in origin, but with some a natural product is chemically converted to the drug we use. Each such chemical structure normally gives rise to several preparations, so that the number of these 'unnatural' substances by no means reflects the extent to which the pharmacopoeia is now synthetic. (The word 'unnatural' is itself arbitrary: chemicals made by a plant and made by a chemist are all part of nature.)

All this traces the pattern of new remedies, but it does not give a picture of their impact. One way of estimating this is the rate of change of the Pharmacopoeia itself. This may be estimated by expressing the new additions to the Pharmacopoeia, at each revision, as a proportion of the total preparations. As well as additions there are deletions, so that there is a sort of steady 'rinsing', which is in fact accompanied by quite a slow growth in total size of the Pharmacopoeia. Figure 14 shows the steady, fairly low rate of revision up to around 1930 and then the accelerating

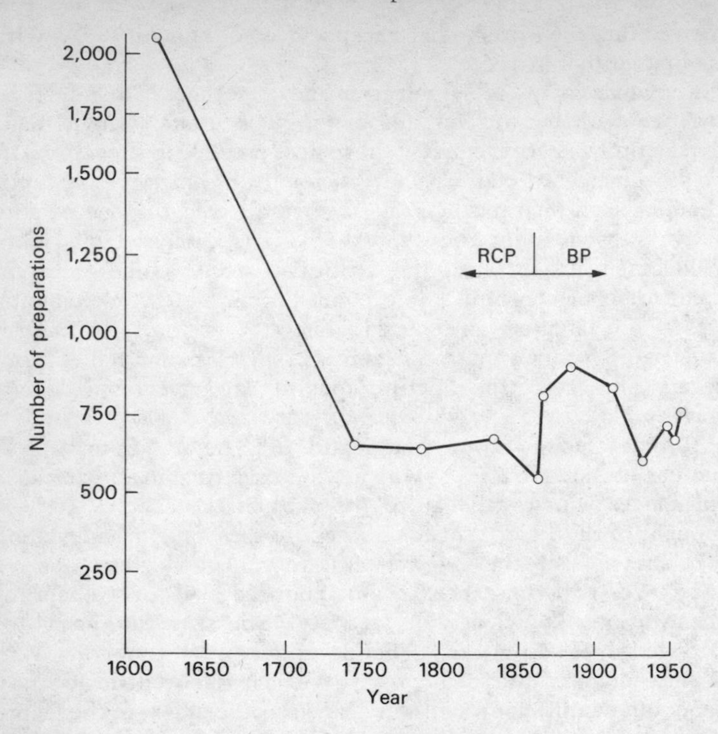

Fig. 12(a). The change in the number of the materia medica and derived preparations in the pharmacopoeias from the first pharmacopoeia of the Royal College of Physicians (RCP) in 1618 to the *British Pharmacopoeia* of 1953 (BP)

From Paton (1979b).

growth up to 1963. At this point, there begins a clear and rapid decline, with the present level of innovation roughly corresponding to that of thirty years ago.

It is in that continuing decline since 1963 that some unease may be felt. The immediate cause was, of course, the thalidomide disaster, resulting in the UK in the formation of the Dunlop Committee, now succeeded by other bodies, which required much more extensive test procedures. But one might have expected that once new regulatory procedures had been established, growth might start again, though from a lower level because of the greater cost of testing. Yet that has not happened. Various explanations

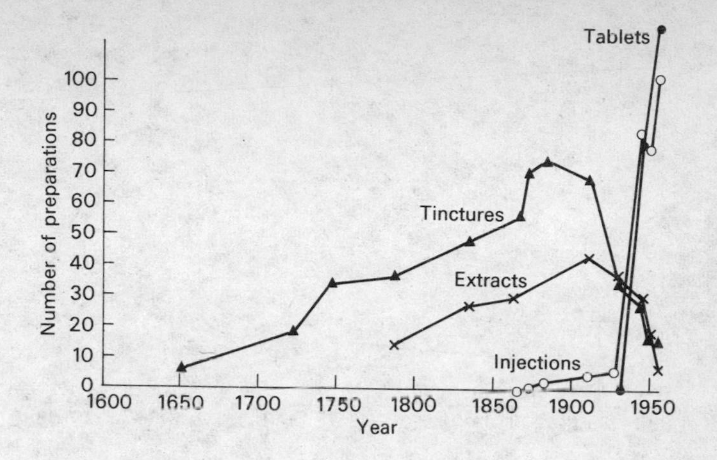

Fig. 12(b). The rise and fall of the number of tinctures and extracts, followed by the rise of tablets and injections, as pure substances become available

From Paton (1979b).

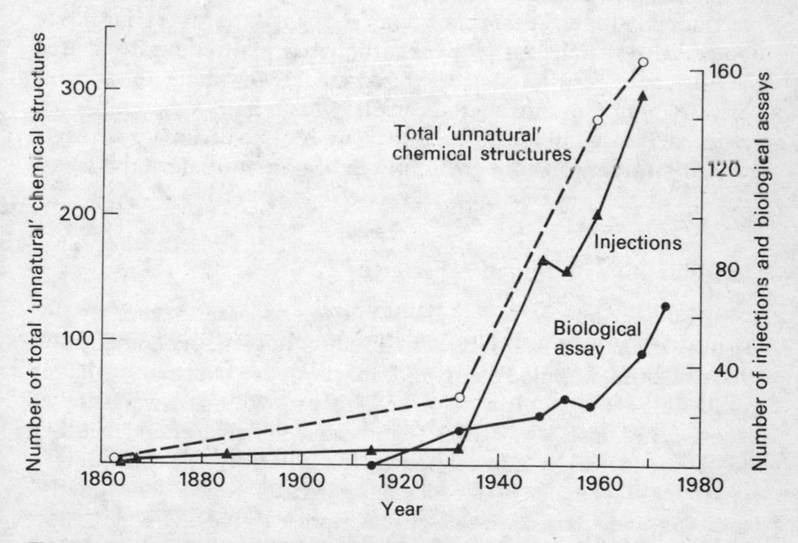

Fig. 13. The growth of the number of injections, of synthetic preparations, and of biological assay in the pharmacopoeias

From Paton (1979b).

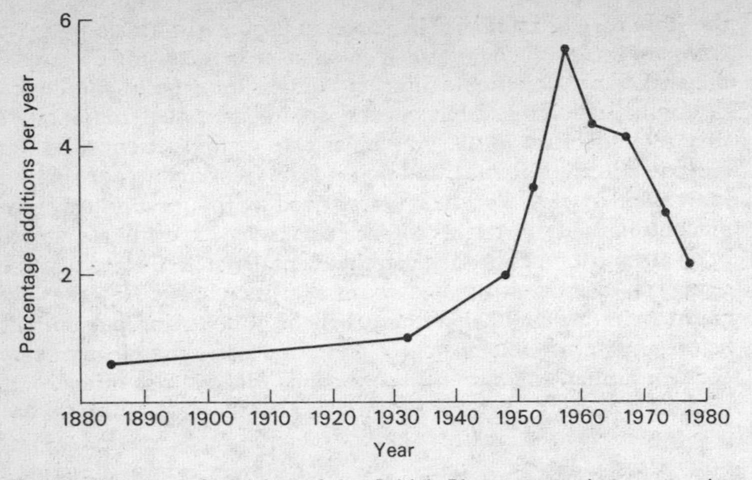

Fig. 14. The rate of revision of the *British Pharmacopoeia* expressed as the percentage new additions each year since the last revision
From Paton (1979b).

may be suggested, and all constitute theories of one sort or another. But one possibility deserves some thought: that the nature of biomedical research is no longer sufficiently centred on integrated systems; as a result too little research is now done in circumstances where the unexpected can show us what we had not even imagined. It is by no means certain, if we look to future knowledge, that our strategy with its great growth of non-animal methods is a wise one.

The alternative of human experiment

In earlier sections we have already discussed reasons for distinguishing man from animals, which led us to consider some of the ethics of human experiment, in particular the suggestion that it would be morally better to use a human who is immature or defective in some way rather than some higher animal. Another and older suggestion, sometimes with a vindictive flavour to it, is that it would be better for the experiments to be done on the experimenters themselves. Such suggestions give the impression that experiments are never done on humans, and that the idea of human experimentation is a novel one. In fact it has a long history, although that history has never been adequately written.[11] Part of the story may be found, for instance, in Pappworth's book, *Human Guinea Pigs*, a trenchant attack on what he believes to be a largely unjustified use of patients or other humans for experimental work in clinical medicine.[12] His general argument is that the doctor's first responsibility is the individual care of the patient who has entrusted himself to him, and that it is not justified to put such an individual at risk for the general good without very stringent precautions and thoroughly informed consent. Experiments should in no circumstances be made on the mentally sick, the aged, or the dying, or on prison inmates or on the 'incurable'. He takes for granted the principle that there will be full testing on animals before any human experiment is done, and notes that nearly every code for human experiment recommends this. He also offers some reasons which explain why animal experiment is sometimes by-passed: namely that animals by law must be anaesthetized for any other than minor procedures; and that human experiments need less form-filling and record-keeping and are not made the subject of an annual return to the Home Office. He accepts, however, that some exception needs to be made for research into mental disease, where the analogous condition does not exist in animals. The growth of ethical committees governing clinical experimentation has gone some way to meet such criticism; but the question of patient's consent, and of what is justifiable, remains actively debated. In that debate, many of the same criticisms that are levelled at animal experiment reappear: that modern medicine is losing a holistic approach, is too 'scientific' and 'analytic', neglects preventive medicine, has been much less successful than it claims to be or is now actually harmful, corrupts its practitioners, and neglects patients' 'rights'. The attack on animal experiment and that on human experiment thus converge to a general attack on winning new knowledge at all, for no one has discovered a better method of doing this than by the 'way of experiment'. In fact, human experiment is like animal experiment—it has its place to improve understanding and to win practical benefit, and the question is of achieving the right balance. But that issue is beyond our present purpose. It is, however, worth surveying something of the pattern of human experiment, partly to indicate its history and its scale, partly to bring out the areas where it can be seen to be especially useful, not indeed as an 'alternative' but as a complement.

Human experiment goes back, of course, to the dawn of history. The vast bulk of a man's knowledge of traditional remedies came from trying out on himself or on others the materials in his surroundings. But so long as there was little knowledge of how the body worked, only the most elementary observations could be made. The doctrine of the humours, with its categories of dry and moist, hot and cold, earth, air, fire and water, yellow and black bile, blood, and phlegm, reads strangely now.[13] Yet it must have seemed very cogent in its day; for all that could be observed would be largely changes in the 'humours': diarrhoea and changes in colour of stools, change in flow and colour of urine, vomiting, sweating, the production of 'phlegm' from lungs and nasopharynx, and bleeding, together with feeling hot or cold, the relief of pain, and gross behavioural change. But once knowledge of the bodily processes, both in normal function and in disease, began to grow, such observations became recognized increasingly as only superficial signs of deeper mechanisms.

The trial of drugs in man continues to this day. Sometimes the chemist or pharmacologist is simply curious to experience what the substance concerned would do to himself, or he may feel that if others may receive it he should receive it first. Serturner, who first isolated the alkaloid morphine from opium, gives a graphic description of its effect on himself and some students. The Czechoslovak Purkinje nearly killed himself with camphor, and gives a fascinating account of the visual effects of a high dose of digitalis. Christison, in Edinburgh, pioneered the study of eserine, the active principle of the Calabar bean used as an African ordeal poison to test for innocence or guilt.[14] The work proved very important and led to the substances known as anticholinesterases, from which came a treatment for eye disease, the first means of relieving myasthenia gravis, a whole generation of insecticides used in every garden, and (in nature's customary two-edged way) the 'nerve gases'. Christison took too large a dose, realized his life was threatened, but was just able to save himself by drinking his shaving water to produce vomiting. When the arrow-poison curare was introduced into surgery, to facilitate operative manoeuvres by relaxing the patient's muscles, a number of anaesthetists and pharmacologists tried it on themselves. One of these experiments, whose subject was S. L. Smith of Salt Lake City, could claim to be among the most courageous experiments ever done.[15] The reason for doing it was that while it was known that curare penetrated only very slowly into the brain, yet it was also known from animal work that if applied directly to the brain, by-passing the 'blood-brain barrier', it caused convulsions. It might therefore be possible that with large doses, or during a long operation, effects on the brain would appear. The experiment was, therefore, to give as high a dose as possible to a conscious subject, using artificial respiration of course. The curare progressively paralyses the subject, so a signalling system was arranged whereby movement of any muscle meant 'Yes', and not moving it meant 'No': it would thus be possible to continue until there was no response, either because the last muscle had been paralysed, or because unconsciousness had occurred. In the event, the last muscle to go was the outer part of the left eyebrow; speech had gone long since, and a dose about three times more than sufficient to paralyse breathing had been reached. The essential observation was made after the subject's rather prolonged recovery; this was that he had been conscious throughout and, indeed, recalled things that the observers had failed to record. If there was any effect on the brain, therefore, it did not touch consciousness, memory, or intelligence.

This last experiment points to one area where human work offers a unique advantage—namely that the subject's testimony can be used. Therefore, whenever functions such as sleep, memory, mood, 'drive', cognitive capacity, or relief of pain are involved, human study is obligatory, as it is for research into mental disease. There has been a great deal of human work on pain and sensation. A great neurologist, Henry Head, and a great surgeon, Wilfred Trotter, both became interested in recovery of sensation after nerve injury, and had nerves in their own arms severed, so as to observe the subsequent changes themselves.[16] A physician, Sir Thomas Lewis, a pioneer of what he termed 'clinical science', explored widely the responses of skin to painful stimuli.[17] He mapped out the characteristic 'triple response' that is so familiar with nettle sting (a dusky reddening—'erythema'—at the point of injury, forming a white 'weal' as tissue fluid accumulates and makes it swell, together with a patchy and pinker reddening in the regions around—the 'flare'), and showed how local injury (such as a hearty slap on the back for sensitive subjects) produced the effect by releasing histamine from cells in the skin. His later book, *Pain*, remains one of the best accounts of its clinical study, and also shows what the physician can learn from a patient's report about it.[18] Another important study was that by C. A. Keele and Desiree Armstrong, who opened up the study of what it is, in injured or diseased tissues, that actually generates the pain sensation.[19] They removed the outer layers of the skin by raising a blister on their own or colleagues' arms with a small mustard plaster, and then cut away the top of the blister. This left the blister base, with pain-sensitive nerve endings exposed, on which they tested various chemicals. The subjects were given a bag to squeeze which recorded their sensations on a scale from zero to very severe. They were able to show that a considerable range of natural substances, liable to appear in inflammation, were potent and rapid in causing pain, graded in intensity with the amount of substance applied. Other techniques used in man have been the 'hot spot', where a patch on the forehead is blackened and light shone on it until pain is elicited, giving a quantitative measure in terms of the number of calories of heat required; or 'ischaemic' pain, where the blood flow to the arm is prevented by a blown-up cuff around the biceps, and then work is done with the forearm until pain develops; or with electric currents to fingers or to the stoppings of teeth, and so on. With such methods new analgesics have been tested, or various aspects of the physiology of pain analysed.[20] Electrical records have been taken from nerves in the wrist as pain is being elicited, and nerves stimulated to characterize the properties of the fibres that actually mediate pain. Such work naturally runs alongside clinical study. One potentially fruitful field is among those who suffer some persistent pain, for they can report not only on relief but also on how long it lasts—an essential datum for clinical use. A patient may welcome such an experiment, since his pain is being put to some use.

A second area of special study has been of the heart and circulation and of the respiration. Man is unusual in his upright posture, so that reflex mechanisms have been developed to stop all his blood pooling in his legs and lower abdomen when he stands up after lying down. Momentary fainting on standing up rapidly after a hot bath, or the occasional collapse of guardsmen on parade, show the control system being stretched to the limit. These 'postural reflexes' can only be studied in humans (although some investigations have been attempted in the giraffe!). So there is a large body of very fruitful work with procedures that lower blood pressure (such as blood-letting, suction on the lower part of the body, or emotional stress), or raise it (adrenaline and related drugs, or pain), and their effects on blood-flow through various parts of the body. All this knowledge now contributes to the treatment of cardiovascular disease, shock, and hypertension; and our knowledge of the postural reflexes has allowed the development of another technique in surgery, that of the bloodless operative field, obtained by using a suitable posture at operation to drain the blood away from the region operated on.

The human study of the control of respiration had a different impetus, namely that the human subject can provide skilled co-operation in complicated experimental procedures. In the days when it was possible only with great difficulty to analyse the blood gases, J. S. Haldane established that by a special respiratory manoeuvre one could obtain a sample of air from the deepest part of the lung, where exchange between blood and air breathed takes place, which provided a measure of the amounts of oxygen and carbon dioxide in the blood. He was then able to show that respiration was largely controlled by keeping the CO_2 level constant; when the body does work and oxygen is used, CO_2 is produced which raises the CO_2 level in the blood, and the breathing increases to wash it out. Subsequently this model proved to be too simple; other factors were identified, and their interaction mapped in a detailed mathematical model. Virtually all the work was done by physiologists on themselves.

A similar participation has helped to throw light on the control of limb and eye movement. Here the subject can try to perform various tasks—such as a maximal muscle contraction, or tracking a target, or responding to a stimulus—during which movement, or new activity, or electric potentials on the globe of the eye or from the skull are recorded.

A third area, which received particular stimulus in the World Wars—particularly the last war—was research to ensure man's safety in the air, under the water, on the surface of the water after shipwreck, or survival in adverse circumstances.[21] For many of the problems, although not all, there was a background of general knowledge of body mechanisms often gained from animal work; but to know precisely how little oxygen, water, or food, or how high a temperature a man could tolerate and still be effective required direct experiment. So the aviation physiologists discovered how high men could fly both without oxygen and with oxygen provided, and how to avoid 'aviation bends' when the pressure was very low, and the effects of the sudden bursting of a pressurized cabin. They defined what acceleration the body could tolerate when coming out of a dive, or if suddenly ejected, or on crashing, and they designed and tested equipment to increase the tolerance. Such work now underpins the safety precautions of a modern airline, the ejector seat, and the lifebelts for survival at sea. Another especially courageous experiment was that of a professor of anaesthetics, E. A. Pask, who wished to verify that a buoyancy suit for survival of a pilot shot down at sea would be sure to keep his face out of the water even if limp and unconscious; so he had himself anaesthetized and given curare to ensure his limpness, and then thrown into a swimming bath wearing the equipment concerned.

The diving and naval physiologists on their side discovered man's tolerance to carbon dioxide and high oxygen pressures: they worked out how to minimize the risk of bends with new depths and underwater tasks, explored how a man could be brought rapidly but safely to the surface from a sunken submarine, and discovered what temperatures he could tolerate and how best to save him from drowning and from death by cold while awaiting rescue. Other research discovered the tolerance to high temperatures and humidity at various rates of work. Related also to civilian life were studies, in which conscientious objectors also often took part, to discover what was the minimum amount of food, water, or various vitamins required for health: this produced fundamental nutritional knowledge in the process.

Chemical warfare had been experienced in the first war, and it was one of the good features of the second not only that mustard gas was not used but that an antidote to it was found. Human experience was already available. A new way of neutralizing such compounds was discovered, and shown to be effective in animals. The investigators then tested it on their own skins, showing that if applied at once the new remedy (BAL, British Anti-Lewisite, dimercaprol) could virtually abolish the effects; but that with time the damage became irreversible, treatment being almost ineffective after twenty-four hours.

Today probably the widest field of human experiment is in clinical pharmacology. As techniques have advanced and medical standards risen, greater and greater attention has been directed to adverse effects in patients and to make treatment more precise. This in turn needs thorough knowledge of the amounts of each drug in the body fluids in relation to dose, the speed of its elimination (for if it is persistent, daily doses will lead to steady accumulation), whether it forms by-products with undesirable effects, and how far these vary with age or disease or genetic inheritance. In addition, far more accurate assessment is made of what a drug does in all its actions and how it interacts with other drugs. While mechanisms may have been mapped out in the test tube, tissue culture, or animal, the final establishing of detailed behaviour now needs to be done in the species in which the drug is to be used—man, horse, cow, sheep, dog, cat, or whatever it may be. So far as man is concerned, then, there has been an explosion of experimental work. If one takes a single issue of a monthly journal of clinical pharmacology, one will find scores of experiments in patients and a similar number on normal volunteers. By contrast, in a journal of general experimental pharmacology, around half or more of the papers will be on isolated tissues, extracts, or enzymes; and the others either on wholly anaesthetized animals, or a few involving conscious animals in work on some central nervous problem or on animals subjected to some drug treatment and then shortly afterwards anaesthetized and killed and the tissues sampled.

Some readers may find this account of experiments by scientists on themselves and on other humans irrelevant to the issue of animal experiment; they may not care what scientists do to each other so long as they leave animals alone. They might note that only *some* experimenters do experiments on themselves, and ask 'If some do this, why not all?' They might conclude that scientists are just a cruel lot, as callous about man as about animals, and that it should all be stopped. They may deplore human experiment under the stimulus of war, or for anything connected with industrial profit. To meet such comments would be to go over old ground. More important, the purpose of outlining the scope of human experiment was rather for the reasons indicated earlier. First it shows that any belief that biomedical scientists as a whole never put themselves, but only animals, at risk is false. Secondly, and more constructively, one can see how human and animal experiment fit together. It is not *only* that man is more valuable than animals, it is also that each species, man included, is particularly appropriate to tackle particular questions: and when experiments can be done only in man, man is used. The pattern of work, in short, is now a dynamic mixture of non-animal, animal, and human experiment, with the choice made according to the investigative need.

Summary

1. The risk of the unnecessary repetition of experiments, and the forces minimizing it, are reviewed. The risk was found to be small by the Littlewood Committee, and available methods of data retrieval now available lead to the same conclusion.

2. The long history is surveyed of the development by the animal experimenter of methods using not whole animals but isolated organs, isolated tissues, tissue culture, isolated cells,

subcellular components, modelling, and structure–activity relationships. Evidence from the Home Office reports, from Medical Research Council reports, and from the scientific literature, all points to the extent to which such methods are used. Study of the revolution of therapeutics and of the recent decline in rate of drug innovation raises the question whether the trend away from experiment on the whole organism may have gone too far.

3. Since it is sometimes implied that there is little experiment on man, the extensive use of the 'human alternative' is also reviewed.

4. It is argued that the correct approach for biomedical research should be a rational mixture of non-animal, animal, and human experiment, as appropriate to the investigation concerned.

9

MAN AND MOUSE

It is sometimes contended that only work on humans can be relevant to human knowledge or benefit, because of the biological differences between men and animals. If we took this view, all medical research should be done on man and all veterinary research on animals, with neither able to benefit from the other. It seems illogical to claim simultaneously that man is so like animals that to distinguish between the two is like sexism or racism, and that man is so unlike animals that research on them is irrelevant to human biology. Much more important, however, is the fact that if the argument in Chapter 2 is accepted, then this logic-chopping is beside the point: man shares much of his biology with animals, but yet has established, by his capacity to accumulate and build on his experience, a qualitative difference. The real issue is that of the respects where their biology is close and where divergent.

One obvious point of divergence is in the area of the highest nervous functions. The observation that man and mouse cells can be fused to form viable hybrids shows that the whole intracellular machinery is compatible at this level. Yet it requires only a glance at a microscope slide or neurological atlas of mouse, rat, dog, monkey, and human brains to see that while elementary mechanisms could be studied in any of them, the more complicated functions can be assessed only on the more highly organized structures. As so often, the two approaches are complementary. Some remarkably primitive systems have been invaluable in mapping out elementary mechanisms, simply by virtue of their relative simplicity and accessibility; examples are groups of nerve cells in the leech or the sea slug, or the nerves to crayfish or lobster muscle, or the junction between crayfish nerve fibres, or the famous squid axon. From these, various patterns of the way excitation and inhibition are transmitted within a cell or from one cell to another can be identified, and tests for characterizing them found, which can then be applied in more complex situations. But when one comes to the problems of how activity from scores of different sources is integrated, the higher systems must be used.

A second major field of difference arises from what is called 'metabolism', the way an organism obtains energy from its food, deals with foreign substances, eliminates waste products, or responds with changes of heat production to environmental change. There is a general rule that one needs a dose of a drug about fifteen times higher in terms of body weight in a mouse than in a man to produce comparable effects.[1] How then can the two be regarded as in any way comparable? The answer is a quite simple one and rather interesting generally, although it may not be familiar. It may be remembered from schooldays that the volume of a sphere is $\frac{4}{3}\pi r^3$, and its surface area $4\pi r^2$. It follows that if one doubled the radius of some specimen sphere to obtain a larger one, then its volume (and therefore weight) would increase by $2 \times 2 \times 2 = 8$, while its surface area would increase only by $2 \times 2 = 4$. In other words, as it becomes bigger, *its surface area would diminish proportionately to its volume.* Conversely, smaller spheres will have a proportionately higher surface area. The same principle operates with shapes other than spherical. This is, of course, why, if you want to dissolve some solid substance in water, it will dissolve much faster if ground into a fine powder, than if left as a large lump or crystal: the weight of the material is the same, but the surface area, at which the solution into water takes place, is enormously increased. Turn now to comparing a mouse with ourselves—both warm-blooded animals, but the mouse having about fifteen times greater surface area in relation to its weight. Because we are warm-blooded we are continually losing heat from our bodies through the body surface (by radiation, convection, and evaporation of water). Consequently the mouse is losing it proportionately much faster than we do, and so needs more food in relation to weight than we do, and has to burn it faster. All its body processes are accelerated, and foreign substances are eliminated faster. This may well contribute to the shorter lifespan. A typical result is that any drug which is broken down in the body and requires some time to act is bound to be apparently proportionately less active the smaller the animal it is tested on. But what is happening is revealed if you actually measure the amount present in the blood: and there you find that for equal effects the blood levels are similar, despite the higher doses in the smaller animal. It is simply that with the higher turnover in the smaller animal, more must be given to obtain a given level. One striking example arises if one is examining the way alcohol is dealt with in the body: an alcohol level that could be disposed of in man in around 4–6 hours would take 8–12 hours in a horse,[2] 2–3 hours in a dog, and ½–1 hour in a mouse. As a result, if you try to make mice drunk, it is quite remarkably difficult! You can give them 20 per cent alcohol (about equivalent to port) as their sole source of water, which they then drink during their active period overnight; yet you will fail to detect any alcohol in their blood in the morning. Though man and mouse differ in this way, therefore, it is a difference one understands and can correct for.

Beyond this, however, there are many individual differences between men and animals: given what was said earlier—that each of the different species stands not in a genetic continuum but each at the end of their own evolutionary chain—this is not surprising, particularly if one also considers differences in diet, sexual cycle, and 'life-style'. In fact, these differences are probably helpful in the end rather than the reverse, provided they are understood. Man's biology varies considerably: as a vet once remarked to me, 'Compared to the average laboratory animal, man is an appalling hybrid!' Some examples in man are the especial liability to sickle cell anaemia and thalassaemia in the African and Mediterranean area; the low acetaldehyde dehydrogenase of the Japanese that makes them flush more readily with alcohol; the latent myasthenic; the subject lacking the enzyme non-specific cholinesterase in the blood, making them especially sensitive to a particular muscle-paralysing agent; the subjects who are 'slow acetylators', who inactivate certain drugs more slowly than the rest of us. The similar variation among animals can be helpful, for it provides a chance of finding an animal model that is comparable for any particular human idiosyncrasy. An interesting case involves vitamin C, where man, in his need for it in his diet, resembles a few animals (such as the guinea-pig and an Indian bird, the red-vented bulbul), but differs from the great majority. The guinea-pig has thus been of great importance for nutritional studies. But it does not stop there, for when one considers the *function* of vitamin C, and its requirement in diseased or stressful states, it is striking to find, in those animals like the rat that can make it for themselves, what large amounts they can manufacture; and this has stimulated interest in the use of vitamin C at doses higher than those required just to prevent scurvy in normal adults. It illustrates yet again how biological information does not exist in atomic fragments, all isolated from each other, but forms a fascinating whole.

But one should not overestimate these differences. We have only to look at an account of evolution, or at textbooks in comparative anatomy, to see how much we have in common: hearts, lungs, kidneys, brains, endocrine glands, nerves, muscles, digestive systems, all built on the same plan. This homology goes back still further as one moves down to the biological elements like the nucleus, mitochondrion, or the cell membrane, out of which the higher organisms are built. One result is that if one opens the British Pharmacopoeia at any point, and asks, 'Is the type of action of this drug the same in an animal as in man?', the answer in all my own trials has always been 'Yes'. The only differences that appear are in dose required, duration of action, sometimes in the way the action manifests itself, and sometimes in side-effects. The same conclusion follows if one simply compares the drugs used by the doctor and the vet: the overlap, as explained in Chapter 5, is very great indeed.

One last example may be quoted. It has been said that if penicillin had been tested on a guinea-pig it would never have been introduced into medicine, because it is so toxic to that species. There is no historical basis for this conclusion.[3] In Fleming's early work penicillin showed simultaneously a remarkable potency against certain bacteria and an equally remarkable freedom from toxicity on white blood cells compared to any previously known substance. Florey and his colleagues found the same: they also discovered that it was harmless in tissue cultures and to mice and rabbits. The higher sensitivity of the guinea-pig to its toxic action was known early on, and there is no evidence that it hampered development in any way. The real difficulties in the early stages were obtaining enough of the material, the painfulness of the injection with the still impure material, and the febrile reactions ('pyrogenic response') also due to impurities. But the guinea-pig result could be regarded in a different way. The first major drawback with penicillin arose when some patients died from allergy to it. Today, therefore, penicillin-sensitivity is always tested before treatment starts. It is interesting, however, that of all animals, the guinea-pig so far seems closest to man, not only in requiring vitamin C but also in its allergic responses and the way these manifest themselves, particularly in giving a bronchospasm comparable to an asthmatic attack in man. The effect of penicillin in the guinea-pig, which has still, to my knowledge, never been fully analysed, could well have provided a warning that allergy to penicillin would appear. Toxic effects tend to be regarded simply as adverse phenomena of no further significance; in fact they are often of considerable significance as pointers to unexpected possibilities.

Summary

Man and mouse and other animals, therefore, have an immense amount of biology in common. Some differences can be recognized from obvious causes, such as varying extent of development of the nervous system, different metabolic rates, different diets and dietary needs, different sexual cycles. When other differences arise they are themselves of considerable interest and can open new possibilities. That may make extrapolation from animals to man, or (for veterinary work) from animal to animal more difficult; but it is also part of the seed-corn of future discovery.

10

STRIKING THE BALANCE

We come, at last, therefore, to the way in which the balance is to be struck, between benefit reaching into the future on the one hand, and the suffering that may be entailed on the other. From what has already been said, we could focus on three requirements:

(a) that if experiments are to be done, they must be 'good science', well designed, tackling significant questions, shaped so as to produce definite answers. If science is not good, animals are wasted.
(b) that the suffering is to be taken into account, and minimized.
(c) that the balance is to be struck responsibly, with some means of assurance for the public that this is so.

In this chapter, therefore, some account will be given of what happens in practice as regards the experimental procedures and the numbers of animals used. This will refer to the situation at the time of writing, although practice is continually changing.

(a) Good science

Science is essentially a social activity: its purpose is not merely the acquiring of scientific knowledge, but also its communication. Some investigators welcome early discussion and criticism of work in progress; others prefer to wait until a fairly full 'story' can be told. All have depended on the communication of their work by earlier or contemporary investigators. If the work is neither discussed while in progress nor reported when complete it is as though it had not been done—except perhaps for some insight or knowledge gained by the observer, which may assist his own later work. One may think of the scientist as a member of the international community in his own day, as well as of a community reaching back into the past and forward into the future; or one may view him simply as a man intent on his own aggrandisement. Either way, the fruit of his work comes only with its communication.

That communication comes essentially at two points: the first when he is planning and seeking resources for the work; the other when he reports his results. At each point it is subject to intensive criticism of the same general type, whether he is submitting a proposal for a grant, giving a seminar to his colleagues, giving a paper to a learned society, or submitting a paper for publication. It will rapidly be pointed out to him if similar work has already been done, or if, from existing knowledge, the outcome is already obvious or trivial. Any defects in technique, in assumptions, or in logic are seized on. The fact that others are already engaged in similar work (which, if not yet reported even in a preliminary form, may not be known to him) is commonly indicated by grant-giving bodies and can lead to a useful collaboration. If the question asked is scientifically insignificant compared to the resources sought, that is quite a common reason for rejection of a grant application. Equally, rejection is common when a good question is asked, but the experiments planned can be seen to be unlikely to lead to a decisive answer. Both in the giving of grants and in the adjudication of papers for publication, the method is that of the so-called 'peer review', in which the opinion is sought of those familiar with the scientific issues and the technical difficulties.

Judgement of this sort can be made only by other scientists. Laymen may look at the titles of published papers, and find them strange, comic, and apparently trivial. Such titles may even draw the attention of legislators and be scoffed at in the cause of reducing scientific expenditure. What effect that has on the custodians of national finances is not clear; but it is obvious that it is no more reasonable to judge a paper merely by its title and without the scientific knowledge required, than to judge a piece of legislation by *its* title and without any knowledge of the situation for which it was framed. Some cases of the misunderstandings that can arise are outlined elsewhere (p. 152–56). The difficulty in correct appreciation of a scientific project is not, however, restricted to laymen. It occurs also among scientists, some of whom have distinct views about the especial use of particular approaches, and the worthlessness of others. For instance, there is a long tradition of misunderstanding between the clinical and the pre-clinical worker. The former may see experiments, ostensibly related to medicine, conducted on frog muscle or crustacean nerve, and may find it hard to see any medical relevance. The pre-clinician, on his side, may see the clinician doing experiments apparently superficial in character and falling short of the rigour he demands. Both are usually wrong, and it is one of the merits of the blurring of the clinical boundary that such misunderstandings are now less common.

One area where experiments with a similar objective, and hence some apparent redundancy, may occur is when some exciting discovery opens a new field—often a technical discovery such as the use of radioisotopes and the scintillation counter, or of a new way to record nervous activity, or of a new chemical technique allowing far more sensitive analyses of important hormones or neurotransmitters. Then a 'race' may start. But it is a general rule that if you put two scientists at the same starting point, they rarely follow the same experimental paths. Nor would one expect them to, since they all differ in the training, reading, and past experience that they bring to bear. Consequently the outcome is often a joint or nearly joint discovery, but reached in different ways, and revealing different aspects. The outcome in fact is a sort of simultaneous confirmation, of the type that is in any case essential for an important finding.

It cannot be pretended that scientific investigation, any more than any other human activity, could not be better done. But those who read the scientific journals or who have attended scientific meetings, will probably agree that the critique offered to the experimenter on scientific grounds is as severe as any that is encountered elsewhere in life. The giving of your first scientific paper to a learned society is in the nature of a 'rite of passage' or a tribal ordeal; and it will be found that seniority does not diminish the awareness of that critical scrutiny. One can fairly claim that the forces maintaining the scientific standard of the animal experimenter are as powerful as those maintaining standards in any other walk of life. There are few other areas in which such continuous, open, and explicit criticism of one's work is provided.

(b) Taking into account and minimizing suffering

Experimental numbers

At this point it is worth turning to the figures provided by the Home Office Annual Report for 1982 with its twenty-one tables of analysis, to consider their implications. First, we may review the trend in use of animals in recent years. Table 9 gives figures since 1971: it will be seen that use levelled off between 1971—the highest point—and 1976, and has then steadily declined, so that in 1982 it was 25 per cent lower than in 1971.

Table 9. UK Home Office returns of animal experiments

Year	Number of experiments (thousands)
1971	5607.4
1972	5327.1
1973	5363.6
1974	5561.2
1975	5379.1
1976	5474.7
1977	5385.6
1978	5195.4
1979	4719.9
1980	4579.5
1981	4344.8
1982	4221.8

Next one may list the number of experiments done (each animal counts as one experiment) according to species used, arranged not zoologically but by number of experiments, together with the percentage of the total (Table 10).

The first point this shows is that, generally speaking, the 'higher' the animal the fewer experiments are carried out. The great majority are done on mice and rats. Fish come relatively high, because of their use in work on environmental pollution, and on 'infection and immunology'. The frog is the main exception, being used in limited numbers for the obvious reason that it is cold-blooded and differs rather substantially from the mammal. Cats, dogs, primates, horses, donkeys, and cross-breeds are each below 0.2 per cent of the total. 'Other ungulates' (hoofed animals), which includes cattle, are involved primarily in veterinary study of normal and abnormal body structure or function and for the development of veterinary products and appliances.

The general pattern of animal use will probably be found reasonable, but the question will still be asked, 'Why so many as 4¼ million?' The answer can only be that that is the number that has been found necessary. The forces working against the unnecessary use of animals, and the increasing scale of the use of alternatives, have already been reviewed—as have some of the benefits that have been won in the past. One can see no basis for deciding from first principles what is too large or what is too small. The current figures represent, very roughly, one mouse per head of

Table 10. Number of experiments performed in Great Britain classified by species

Species	Number	Percentage of total
Mouse	2,442,702	58
Rat	932,335	22
Bird	251,818	6.0
Fish	165,833	3.9
Rabbit	164,993	3.9
Guinea-pig	154,740	3.7
Other rodents	33,785	0.8
Ungulates (other than horse, donkey, cross-breed	33,574	0.8
Dog	13,146	0.3
Reptile or amphibian	7,882	0.18
Cat	7,341	0.17
Primate	5,654	0.13
Other mammal	4,583	0.11
Equidae (horse, donkey, cross-breed)	475	0.01
Total	4,221,801	

the population every ten years: is that big or small, in relation to the animals used for food each year (in the hundreds of millions), or dying naturally, or being destroyed as unwanted strays? One cannot even feel that such comparisons are satisfactory. If something is wrong, it is not made any better if something else is worse; that only means we should assign more priority to dealing with what we regard as the worse. One point may be useful, however: the aggregate figure of experiments done is somewhat misleading, since it is built up from a large number of different categories of work. Tables 12 and 13 (pp. 138 and 139) show some of these categories, from which a better judgement can be made as to the justifiability of the various numbers.

A comparison with the numbers of observations needed in areas of enquiry more familiar to the layman may also help, such as opinion or consumer surveys. Few would attach much weight to a survey of, say, 100 people for their opinion on union activity, nuclear waste, or the siting of a new road, as representing a true measure of the opinion of the country at large. Experimental work, because of the variability of living organisms, equally apparent in human political views and in simpler biological responses, is entirely analogous; and to establish some fact, so that it can be the *basis for future action* is just as demanding of numbers. Indeed if we review the numbers used in estimating a frequency of opinion, developing a soil for a plant, testing an insecticide, analysing the composition of paper in old books, or finding out the potency of an antidiabetic drug, they are of the same order—sometimes small, sometimes large according to the precision required.

Experiments wholly under anaesthesia

These experiments, 4 per cent of the total, must be regarded as involving no more pain or suffering than is involved in the putting down of a domestic pet by the use of an anaesthetic, since the experiments are done wholly under anaesthesia and the animal is then killed. The proportion of such experiments for each species is given in Table 11.

Table 11. Percentage of experiments performed in Great Britain wholly under anaesthesia classified by species

Species	%	Species	%
Cat	73	Other rodent	6.4
Other mammal	61	Rabbit	3.6
Dog	30	Horse, donkey, cross-breed	2.7
Guinea-pig	16	Other ungulates	2.5
Reptile or amphibian	14	Mouse	0.6
Rat	11	Fish	0.6
Primate	8	Bird	0.4

This category of experiment consists largely of analytic experiments in physiology and pharmacology and the experimental procedures used would be wholly unacceptable if the animals were not anaesthetized. The cat and dog owe their dominant place partly to the analogies between their physiology and that of man (particularly as regards heart and circulation, respiratory function, and neuromuscular function) and partly because their size allows experiments that would be impossible in a smaller animal—although as technology improves, work on the anaesthetized rat is becoming more feasible. With animals larger than cat or dog—such as primates, ungulates and equidae—their cost alone makes it necessary to find a way that yields more information than an acute experiment can do.

Experiments without anaesthesia

Under the 1876 Cruelty to Animals Act, experiments without any anaesthesia, 80 per cent of the total, must not involve any operative procedure more severe than simple inoculation or superficial venesection (the human counterpart is the discomfort, if any, involved in a polio 'jab' or a blood donation). Pain might, however, result subsequently, for instance as a result of some drug administration. So these experiments are further subject to the 'pain condition' which states:

(a) If an animal at any time during any of the said experiments is found to be suffering pain which is either severe or is likely to endure, and if the main result of the experiment has been attained, the animal shall forthwith be painlessly killed.
(b) If an animal at any time during any of the said experiments is found to be suffering severe pain which is likely to endure, such animal shall forthwith be painlessly killed.
(c) If an animal appears to an Inspector to be suffering considerable pain, and if such Inspector directs such animal to be destroyed, it shall forthwith be painlessly killed.

This condition has led some to conclude that severe pain is common is experimental work. In fact it is very rare, partly because scientists are as humane as anyone else, partly because if it occurred it would destroy the effectiveness of the experiment through the physiological disturbances it causes. In my own experience I have never seen severe pain inflicted on, or experienced by, an experimental animal, but only on human subjects (see pp. 116–20). (The most severe pain I have ever inflicted on an animal myself was when shampooing the family cocker spaniel; some soap got in its eye and the lachrymation, spasm of orbital muscles, whining, struggling, and final escape from the operator were formidable! Such experience makes the usefulness of some less irritant material very obvious.)

For the majority of experiments in this category the induction of anaesthesia would be more distressing than its omission. But some experiments certainly involve discomfort. For instance, a new analgesic may be tested in various ways. One is to restrain a rat, and to lay its tail across a wire, which is then heated by passing a current through it; the time is measured before the rat flicks its tail away. A second method is to place a mouse or rat on a plate heated to about 60 °C (like a hot cup of tea), and the time is measured before it jumps off. The principle with both procedures is that the animal is free to escape from the pain when it is felt. A second category is in psychological experiment, where, for instance, an electrified wire grid may be used in training an animal in some task. The stimulus, however, only requires to be strong enough to change behaviour, and involves only mild pain. A human analogy would be the discomfort that leads to choosing a more comfortable chair. Severe pain would disrupt the experiment.

A much more serious issue is that of 'inescapable' pain of any duration, which has recently come to the fore as a result of work elsewhere in the world. The experiments entailed, for instance, severe stimuli from an electrified grid for a sustained period: the animal stops trying to escape, and there is some evidence that mechanisms in the brain are activated which produce some measure of analgesia. Subsequently the animal enters a state which could be regarded as a 'model' for some type of depression. Pain of this order certainly occurs in man (for example, causalgia and tic douloureux) and presumably also in animals, and severe depression rather than pain *per se* is arguably the cause of the deepest human suffering. A case for such experiments can therefore be made. But permission has not been given for such experiments in Britain. In my view this is entirely correct according to the Act; and it is doubtful whether the physiological information sought cannot in fact be found in other ways.

Two points arise which are important for future legislation. The first is that the 'pain condition' is restricted to pain *specifically*. This may well have been appropriate in its day, when other forms of suffering were commonplace. But with the general rise in standards, it is generally accepted that the field of operation of such a rule needs extending; and words like 'distress' or 'suffering' are now being included in legislative proposals.

The second point is that the 'pain condition', as it stands, means that *however important* some experiment is, if the pain were severe and likely to endure, the experiment must be terminated whether or not its end has been achieved. It therefore represents an absolute stop to the freedom of human enquiry in that particular direction. This became a matter of considerable debate during the preparation of the draft legislation on animal experiment for the European Parliament, since some countries incorporated into their constitution the requirement that human enquiry must be absolutely free. As it stands, the British pain condition means that, however great or widespread the suffering in man or animal to the relief of which an experiment might lead, that experiment would be prohibited if it entailed inflicting severe *and* enduring suffering on one mouse. This would be the case even in some national emergency, or with the appearance of some new disease. The debate is, so far as one can judge, largely theoretical; but it is significant in reminding us that restraint on animal experiment may involve a price that the public is not willing to pay. In the end, the European Draft Convention has allowed severe and enduring suffering, if it is specifically declared or specifically authorized: authorization shall be refused 'if the procedure is not of exceptional importance for meeting the essential needs of man or animal including the solution of scientific problems'. Nations adhering to the Convention, however, may adopt stricter regulations if their constitution allows. So far as Great Britain is concerned, that appears likely. It also seems probable that if a case of sufficient importance arose in practice (for instance in a national emergency) it could be dealt with by special emergency legislation.

Experiments with anaesthesia for part of the experiment

This category provides for those experiments where first some procedure more severe than inoculation or superficial venesection is done under general anaesthesia; the animal then recovers from the anaesthetic and the experimental observations are made. They constitute 16 per cent of all experiments and are distributed over all species. With both these and with experiments without any anaesthesia, the assurance is required that anaesthesia during the experimental observations would frustrate the purpose of the experiment. This, of course, is self-evident for any behavioural experiment or analgesic test. It is also the case for the whole area of safety evaluation, because all anaesthetics have themselves significant toxic actions, and because anaesthetics could mask in a variety of ways the expression of adverse effects. One type of experiment is to implant a fine tube under anaesthesia so that samples of blood or other body fluid can be taken or lymph drained over a period of time without disturbance. Similarly, a suitable device may be implanted to allow blood-pressure or blood-flow to be monitored, or recordings to be made of muscle, nerve, or brain activity in the unrestrained conscious animal. Another category is where a lesion may be made to study the results of trauma, or the way the body's systems of repair operate, or to throw light on the function of some nerve pathway or group of nerve cells in the brain. Thus a tendon may be severed, a piece of bone removed, a material possibly suitable for hip replacement implanted, a suture material implanted to test for irritation and readiness to be absorbed, a nerve sectioned, a local injection made of some substance which inactivates some nerve pathway, a known number of tumour cells implanted in cancer research, a graft of some tissue or organ made in developing new replacement techniques, part of the liver or kidney or some gland removed. With each of these the suffering entailed is first that of the induction of anaesthesia, next that of the post-operative period (which is astonishingly brief in animals by human standards), and then of any disability that may arise. Commonly, whatever the reaction that is being studied it is not allowed to progress far, partly because it is the early stages that are of most interest, partly because if distress is caused it complicates interpretation very severely. This is an area where advancing technology offers considerable hope, by allowing earlier and earlier changes to be recognized. Experiments of this type are, of course (like those with no anaesthesia at all) subject to the 'pain condition'.

Some special experimental areas

To help clarify the discussion, Table 12 shows the number of experiments in certain categories, where the specific scientific area of experiment can be identified. This helps to put some of the issues in better context. Certain related fields have been grouped together.

Another field in which some quantitative insight can be gained from the statistics is where the experiments are done because of something in the nature of a mandatory obligation. One such case is that of the direct diagnosis of disease, where, for instance, it is necessary for clinical or epidemiological reasons to test some material on animals to identify tuberculosis or other organisms. Work of this sort falls particularly to public health laboratories and hospitals. A much larger field is that of drug development, both in its early stages of identifying suitable products, and in the later stages of meeting regulatory requirements. Table 13 gives some of the information. The distinction is not altogether straightforward, since experimental work takes place over a number of years, and work of subsequent relevance to a regulatory body will often have been done before a final selection of product was made for submission to the regulatory body. This will be followed by work specifically for registration of a product, and in some instances by continuing work for batch control. A third category involves the

Table 12. Some special experimental areas in Great Britain

	Number of experiments	Percentage of total
Infection and/or immunology	1,563,999	37
Acute toxicity	445,723	10.6
Subacute/chronic toxicity	187,481	4.4
Fate of drug in body	136,576	3.2
Teratology	26,821	0.6
Neoplasia (cancer induction, screening, prevention therapy)	179,287	4.2
Use of aversive stimuli	36,881	0.9
Application of substances to the eye	19,124	0.45
Burning or scalding by any means	23	
Hazards and safety evaluation (total) made up of:	203,645	4.8
Substances used in industry	66,185	
Pesticides and herbicides	48,101	
Environmental pollutants	27,333	
Food additives	20,125	
Cosmetics and toiletries	18,864	
Household materials	13,934	
Injurious plants and toxins	5,889	
Tobacco	3,214	

testing required rather generally of any new substance that humans may be exposed to, for instance in a factory or in the fields.

It must be stressed at this point that attempts to add up categories may be very misleading, because there are very many cross-entries. The Home Office Annual Return now requires entry of experiments done by the licensee in the past year, subdivided among 8 lists, which between them contain 72 separate categories. For the most part, it is fairly straightforward, but it will be obvious that in such analyses the same experiments could—in the case of a new antibiotic, for example—appear under 'Selection of products', 'Infection and/or immunology', and 'Fate of drug in body'.

These two tables bring out certain dominant areas of work: infection and immunity; work in developing new products and in satisfying or preparing to satisfy regulatory bodies, and, within that, toxicity testing; cancer research; and safety evaluation. Some commentators have made much of experiments involving burning or scalding, safety evaluation, work on tobacco products, and experiments on the eye. The actual figures are worth comparing with previous claims. In assessing these figures, one may also reflect on the frequency of burning in accidents, or on the incidence of blindness, and on the extent and effects of tobacco smoking.

Table 13. Categories of experiments performed in Great Britain for diagnostic or legislative reasons

	Number of experiments	Percentage of total
Direct diagnosis of disease (medical, veterinary, or dental)	113,969	2.7
Selection of medical, dental or veterinary products	1,591,414	37.7
For registration under Medicines Act or overseas equivalent	132,221	3.1
Batch control under Medicines Act or overseas equivalent	582,554	13.8
Health and Safety and Work Act (1974) Agricultural Poisonous Substances Act (1952) Food and Drugs Act (1955) or equivalent overseas	97,663	2.3
More than one of the above	230,731	5.5
Total	2,839,919	67.3

Cosmetics and toiletries and the Draize test

Among the uses of animals none has aroused deeper concern than their employment for the safety testing of cosmetics. A private member's Bill was introduced into the House of Lords in 1977 specifically to make it illegal. Statements were made that 'millions' of animals were used for such tests, and the impression was given of large numbers of animals being tortured to allow industrial profit out of female vanity. The cosmetics industry remained largely silent, and a good many scientists as well as 'laymen' joined in condemnation of such animal use. Yet such condemnation was not universal, and Baroness Phillips's Bill was not carried in the Lords. Some of the reasons for this become clearer if we look a little more closely at the facts.

There is no clear distinction, in fact, between medicinal and non-medicinal products. One may use a salve to relieve chapped lips, or a cold cream for dry itchy skin in old age. An adolescent may use an antiseptic ointment on acne. Gardeners and housewives apply barrier creams to their hands to stop them becoming rough. Such instances merge imperceptibly into recognizable medical treatment for dermatitis or skin infection. Equally, the use of tinted powders or creams to hide some blemish have their medical parallel in the more elaborate treatment of port-wine stains and other birthmarks. The practical dividing line is whether or not a medical prescription is needed for the skin preparation. It is unfortunate that cosmetics and toiletries have such exclusively frivolous associations; both the dermatologist and the social psychologist know better. So we are concerned, not just with an eye-shadow or two, but with substances used daily by millions, for purposes which may seem minor but can be very important in everyday life, and involving direct application in substantial amounts directly to the body.

The history of cosmetics and toiletries is important. It goes back, of course, to the earliest ages: a wide variety of perfumes, oils, and salves have been used over the centuries. Some pigments were toxic by modern standards: the Egyptians used antimony and lead sulphides for painting their eyelids black, or green with copper-containing minerals. The Romans used white lead on their faces or dusted onto their hair, or took red arsenic for a fair complexion. Such substances are used in various parts of the world to this day, but most modern cosmetics have developed from advances in chemistry. Until about World War II the principal agent used for washing was soap. But then the chemistry of long-chain compounds, that has led us to the modern plastic, began to grow. Early discoveries were first the cationic and then the anionic detergents —artificial soap made by adding the appropriate chemical group to a long paraffin chain. With the anionic detergents, materials of unparalleled cleaning power became available, and the cationic detergents proved to be outstanding antiseptics. But they were also found to be capable of irritating the eyeball. In addition there had previously been several deaths from a depilatory, fifteen to twenty injuries from a shampoo, dermatitis and disfigurement from a hair lacquer, and blindness from an eye mascara. It is very instructive to read the US hearings on the subject, both in 1952[1] and two decades later in 1975[2] when the Eagleton Bill, demanding far more stringent tests, was being considered. One study, in 1944, which hoped to show that the use of a cationic detergent would improve the drug treatment of eye disease by improving penetration of drug into the eye, found instead that it could cause severe eye damage. It is not surprising, therefore, that the US Food and Drug Administration commissioned, through the hands of J. H. Draize and his colleagues,[3] the development of rigorous tests of toxicity to skin and mucous membranes. Their purpose was not confined to toiletries, but covered a range of other chemicals to which exposure might occur, through spillage in a factory for instance. Tests for sensitization were also included.

If a test does not give definite and reliable results, it is not worth having. The Draize eye test met this demand. The eye of the rabbit was chosen, a fairly obvious choice for ease of application and ease of examination of the eye (it is not true that it was chosen because it cannot form tears, or has no nasolacrimal duct: it can and has). A defined procedure was worked out of giving 0.1 millilitres of test material into the conjunctival sac. Observations, based on a clearly defined scale of effect on cornea, iris, or conjuctiva, were made at 1, 24, and 48 hours, and at 96 hours if there was any residual effect. The design of the experiments was arranged to be suitable for statistical test, controlled conditions were arranged, and the actual observer was ignorant of the treatment involved, to avoid bias. Since then practice has been changed according to the substance concerned, and when appropriate the amount instilled is left there only briefly and then washed out. The obvious point is worth making, that for the industrialist seeking to develop a new toiletry the objective is to find compounds that fail to give a positive response, rather than the reverse.

It has been suggested that such experiments should be done only under local anaesthesia. There are three reasons for not doing so. Anaesthesia of the eye is itself an interference which would remove the ability to respond to any casual irritant such as dust or a bit of hair; anaesthetized surfaces are particularly liable to casual damage. Second, the anaesthesia would make it impossible to recognize if the substance produced any discomfort or pain; sensory irritation and tissue damage by no means go in parallel. Third, it is an important physiological fact that the nerves in the skin and conjunctiva themselves participate in the responses to many irritants; and if the eye is anaesthetized, the irritant effect is much reduced below that seen in a normal eye, and any harmfulness would be underestimated.

The eye test, of course, was not the only one developed. Other tests included application on the skin over a long enough time to detect an allergic response; and tests for irritancy to the skin, not just on intact skin but also after lightly abrading the skin or removing the outermost layers by repeated application and removal of sellotape. The latter would give some approximation to the abrasions of everyday life with men or animals. Another test was of how far the material is absorbed into the body from the skin. Tests of this type, with various refinements, continue to be necessary.

Finally, there is the question of safety of the material if, instead of application in the normal way, it is itself accidentally consumed, for instance by a child. Before dismissing the relevance of this, we might consider some figures from a report in 1976 by the Department of Prices and Consumer Protection, on 'Child Poisoning from Household Products'.[4] It was known that in England and Wales in 1967–72 there were 10,500 hospital admissions of children of ages 0–4 for suspected poisoning, and the report estimates a real incidence (allowing for out-patient and practitioner treatment) of 40,000 a year. The survey was conducted to study in more detail the ingestion of household products. This was done by following up statistics from six chosen hospitals. Of 1,723 cases, 567 (34 per cent) involved household products and 554 were followed up. Around 100 different substances were identified, of which turpentine (67 cases) and bleach (50 cases) were the commonest. Among toiletries and cosmetics were perfume (26), nail-polish remover (18), aftershave (9), toilet deodorant (8), hair dye (5), soap products (5), hair shampoo (4), face cream (4), nail polish (2), hair lacquer, face freshener, denture cleanser, hair-setting lotion, and false nail glue solvent (1 each). The principal place of consumption was, as one would expect, the bedroom. It is also reported that one child aged two drank a whole bottle of sherry, and that another aged twenty-two months swallowed an estimated 10 grams of gold chloride powder. Thus, somewhere between 3 and 6 per cent of all such hospital admissions involved toiletries and cosmetics, according to what is counted, giving a national total of 300–600 hospital admissions a year and a larger real incidence.

Cosmetics and toiletries, of their very nature, are generally of very low toxicity, and to produce overt toxic effects in an animal would require enormous dosage. For such products, therefore, a 'limit test' is used, whereby the substance is tested only up to some limited dose; even if it is harmless at that point, higher doses are not tested. The question then is what would be a safe limit to use;

and 5 grams per kilogram is a common choice, on the basis that even in exceptional cases, doses higher than this will not be taken. Is it safe enough or too safe? The survey discussed above showed that 80 per cent of the cases involved children between one and three years old: taking a weight of 12 kilograms as an average, then a child would have to consume about 60 grams or millilitres (or about 4 ounces) of one of these substances, to reach a dose of 5 grams per kilogram. It is a matter for judgement whether that allows sufficient safety factor.

The estimates made by some about the number of cosmetic and toiletry experiments on animals proved grossly misleading—and indeed this was obvious to anyone familiar with biomedical work. But direct evidence had to wait until the Home Office framed the new form for the annual return of experiments done, in which 'cosmetics and toiletries' were separately categorized. We now know that the true number over the last six years has been between 20,000 and 30,000 a year (18,864 in 1982). To this are added around 15,000 experiments a year on human volunteers. Allowing for the numbers of experiments required to reach a reliable conclusion, and the range of kinds of toiletry, these numbers are not large. It is not of course necessary to test every constituent in a new product, but only new substances, or old substances used in a sufficiently new way.

Two comments are often made in this context. The first is that we have enough cosmetics and toiletries anyway, and do not need any more: and looking round a chemist's shop, that seems convincing. But it is worth recalling that that could have been said about soap in the 1930s, when cracked hands and washerwoman's dermatitis were familiar sights. It was only by being able to take advantage of chemical advances that today's detergents are possible, along with all the other improvements in means of care for sensitive or blemished skin. Second, although we might be reluctant to forgo whatever advances in chemistry or dermatology might be made by pharmacology and biochemistry in the future, we could still think that their exploitation in toiletries does not warrant animal use. It would then follow that testing would become completely arbitrary, depending, as it would have to, on the experience of those not satisfied with present materials. Without the sort of organized testing procedures we have at present, our knowledge would be dangerously incomplete. It could also be argued that since the use of toiletries is voluntary, it is the people who use them, not animals, that should take any risk. That would mean, of course, that if new toiletries proved harmful, it would be precisely those for whom present materials were unsatisfactory that would suffer.

No attempt will be made to suggest a conclusion about cosmetics and toiletries, chiefly because of the lack of information about cosmetic and toiletry composition and use. There is, however, one general issue about which people must make up their own minds. It is a fact that such substances are used on a huge scale, by the foolish and careless as well as by the observant and prudent; and this use is voluntary. It may be regarded as not unreasonable to expect the prudent to look after themselves and their families; but is it, or is it not, a proper activity of society to seek to protect the others from their carelessness or foolishness? Is it right to provide care for the attempted suicide, or the rash motor cyclist involved in a crash, or the alcoholic or drug addict? And if these vulnerable individuals are to be provided for in some measure, does that, in the case of toiletries, extend to the sort of animal experiments outlined?

Toxicity testing

A second area of especial concern is that of toxicity testing in general, particularly the 'LD_{50}' test. It is probable that ways will be found to reduce the use both of this test and animal toxicity testing in general; but it is important to understand how the LD_{50} concept came to be introduced, for some other way will frequently be needed to serve the purpose it fulfils. Before scientists came to understand and deal with the statistics of biological variability, attempts were made to estimate quantities such as the 'minimum lethal dose' or the 'maximum tolerated dose' in experiments where evidence of liability to cause damage or death were required. These entailed, as the names implied, trying to find doses of some substance which *just* managed to kill or *just* failed to kill an animal. The fact of biological variation meant that animals varied considerably, and if the test was conducted on a group of, say, ten, sometimes a given dose would kill none, sometimes one or two, occasionally more. One of the purposes of such a test was to determine the potency of an antitoxin for human use; and this was done, for lack of any chemical method, by discovering how much was needed precisely to neutralize a standard quantity of some reference toxin. Another purpose was to test the potency of a tincture of digitalis (extract of foxglove, used in heart disease), since its strength was liable to vary with the various crops;

discovering the lethal dose for a frog was one way of doing this. Quantitative tests of this sort are called 'assays', probably by analogy with the assaying of gold and silver. In these tests the measurement involved a 'yes or no' result, known now technically as a 'quantal assay', involving the *proportion* of a group which responds in some particular way. (A general election may be regarded as a quantal assay of political potency.) It required the insight of J. W. Trevan[5] and later of J. H. Gaddum[6] to see that, for mathematical reasons, the most accurate region to work in with such assays is where the response rate is around 50 per cent, in the general range of 30–70 per cent; and they worked out methods for estimating the potency at the point where accuracy would be highest, namely the dose producing 50 per cent lethality. The LD_{50}, or 'estimated dose lethal in 50 per cent of a group', and the corresponding ED_{50}, or 'estimated dose effective in 50 per cent of a group' then came into general use because it greatly improved the accuracy of the work, so that fewer animals were needed for a given degree of precision. The term does *not* mean that 100 animals must be tested, nor that exactly 50 of them must die. The 50 refers only to the proportion, given as a percentage, and could have been $LD_{0.5}$ as a way of expressing the proportion concerned.

A further point about the LD_{50} is that any particular figure is almost meaningless by itself, for in its very nature it, too, is subject to variation from one test to another. The other great advance involved was in conducting the test in such a way that it also gave an estimate of the *variability* of the response. An LD_{50} figure, therefore, should always be accompanied by another figure or set of figures expressing that variability: either what is known technically as the 'standard deviation of the mean' or 'standard error' (s.e.), written for instance as an LD_{50} of 83 mg/kg ± 10 (s.e.); or by a pair of figures giving the so-called 95 per cent confidence limits—in other words, the figures represent the limits within which it is believed the 'true' figure is to be found with that degree of assurance.

This approach represented a considerable advance, and became widely used. One use was to combine ED_{50} and LD_{50} into a 'therapeutic ratio', LD_{50}/ED_{50}, with which the larger the value, the safer the drug. It came to be regarded as a standard 'datum', and to be called for by regulatory bodies. In the process, investigators and authorities began to lose sight of its original purpose—namely to provide an accurate value *where accuracy was required*—and it became used where a much cruder estimate of toxicity would suffice. It also meant that lethality *per se* tended to become the key sign of toxicity. With highly potent drugs or those used relatively close to their lethal level, there is some reason for this. If one is seeking to identify the safest drug in a series of related ones, it may also be desirable. But apart from these uses, much more valuable information is the *nature* of the toxic effect, and the pattern of its earliest manifestation and subsequent development. It has also become clear that there is appreciable variation in value of LD_{50} estimates when the same substance is tested in different laboratories, arising from factors such as differences in detail of procedure or strain of animal. This variation was large enough to make untenable any idea that the LD_{50} constituted a standard value comparable, say, to a melting point, which could be attached to a particular substance. In addition, experience by the special clinical units dealing with poisoning shows that knowledge of a formal LD_{50} provides little help in management of the cases. As a result of such experience, as well as of the cost of the tests themselves, toxicologists themselves have for many years been exploring other approaches. Some simply seek to reduce the number used. Another approach has been, not to aim for a 50 per cent point, but to approach gradually from sub-lethal doses. A third, the 'limit dose' method already discussed, is to accept that as soon as the dose given becomes large enough to give a sufficient factor of safety in practice, there is no need to test higher doses, even if death or overt toxicity has not been encountered.

A particularly interesting development has been a proposal by the British Toxicology Society for the case where a broad classification of substances under headings such as 'highly toxic', 'hazardous', and 'unclassified' will suffice.[7] Such a classification is needed, for instance, for the safe handling and transport of a very wide range of chemicals. Hitherto, it has been assumed in such tests that *lethality* is an essential objective. But, quite apart from the moral objection to inflicting death without good cause, the real objective of safety evaluation is different, namely how to *avoid* lethality. The proposal, therefore, is to base the test explicitly on the signs of toxicity. If there appears to be severe suffering, the procedure allows the animal to be painlessly killed forthwith without influencing the result. Since the procedure uses the appearance of the signs of toxicity as the basis of classification, it requires more attention during its conduct; but it also yields specific information about the signs of toxic action. A weakness may be that if an estimate of statistical confidence in the classification is required, it may not in fact be available from the test itself, as it would be from a formal LD_{50} test. But there is a good deal of LD_{50} data now available, from which the general pattern of variability of response over a wide range of substances can be gathered. So far, estimates of the outcome of the new proposal applied to the past LD_{50} data seem rather encouraging.

Toxicity testing is outstandingly the area where regulatory or consumer pressure comes into conflict with animal welfare considerations. For substances to which very large numbers of people and animals will be exposed, we cannot expect toxicity testing to be avoided. It is not even possible to say that the LD_{50} test can be safely abolished. But it is becoming clear from the body of toxicological information now available and from experience of its usefulness or otherwise that there is scope for estimates of toxicity of less exactness; that there is a need for more information on the early signs of toxicity rather than for estimates of lethal doses; that there is hope of moving away from lethality as an objective in such tests; and that regulatory bodies are becoming aware of these new possibilities.

(c) The exercise of responsibility

In this chapter we have examined the forces maintaining the *scientific* standard of animal experiment; and we have looked at the actual decisions taken, and the final outcome in terms of animals used as shown by the Home Office returns, so that the reader can make his own judgement about it. But he may still wish to know more about the control mechanisms involved.

The context of animal experiment

First, at the risk of some repetition, it is worth pulling together some of the elements of the context in which the animal experimenter works. Earlier, some of this was outlined in connection with the maintenance of scientific standards; but they operate, too, on the *numbers* of animals used. The impression is sometimes given that the experimenter works in isolation, responsible only to himself, free to do anything he wishes on as many animals as he wishes. Nothing could be further from the truth, and some account of the way experimental work takes place is necessary, especially because that context is important in controlling the number of animals used, and the methods and procedures used.

Before an experiment can be done, a place must be obtained in registered premises. Then stipend, equipment, technical assistance, and running expenses are required. All of these, whether in universities, industry, or research institutes, are in short supply. There is, therefore, competition for these resources, and close scrutiny of all the costs, both in departments with regard to their own budgets, and in grant-giving bodies in relation to their available funds. With the latter, rates of rejection of application for support may range, according to the economic climate, up to 60–70%; and the critique offered by the grant-giving bodies extends as liberally to senior as to junior investigators.

Secondly, a licence must be obtained, together (under present British practice) with such supporting certificates as are needed. To do this, an outline of the work to be performed, the reasons for it, and the species of animals to be used must be given. This has to be supported by a professor in one of a range of biomedical subjects, who may be the head of the department, or, in the case of industry or of such a professor himself, from another department or institution. A second supporting signature from the President of a Royal College or the Royal Society is also required. As the number of licensees has grown, this second signature has with some Colleges become a formality, but this is not always, and need not be, the case. The application is usually discussed at some stage with the local Home Office inspector. Necessary supporting material may include, with new licensees, evidence of scientific standing and of specifically biological experience; or with some new technique, further details. New licensees, in my experience, are usually interviewed by the inspector. With new investigators, technicians, or those from abroad, supervision by an existing senior licensee is normally required.

Given the approval of the application, experimental work may start. It is then subject to visits, without notice, by the local inspector, who also inspects the experimental animal-holding provision, and may talk to animal technicians. At the end of each year, a return must be made to the Home Office of animals used in experimental work, each animal counting as one experiment, with their use categorized under various headings. Reprints of published work involving animal experiment are also required.

That is an outline of the formal procedure. But it must also be stressed that scientists do not work in isolation. They are continually exposed to the opinions of their colleagues, graduate students, students, and technicians. A primary objective of science is that it is shared with others; and while this may be dismissed as the 'publish at all costs' disease, it is in fact at the heart of scientific work. Accordingly, the work done is reported in departmental seminars and to learned societies. For the work to be assessed, it is obligatory to outline the procedures used. If procedures are felt to be inhumane, that may be questioned by members of the society. The work is then written up for publication. I can only speak here of British practice, but here at least, account is taken of the humanity of the experiments done, and publication may be refused to those regarded as inhumane. Finally, the work may appear, and is then subject to scrutiny by welfare societies, who have published analyses of their findings.

Does animal work corrupt the experimenter?

It is sometimes suggested that experimental scientists are bound to become 'desensitized' to the suffering (if any) that they inflict on their experimental animals. Similar remarks are made about the medical profession; and a layman seeing a surgeon calmly dealing with a surgical emergency might feel that he, too, was unfeeling and cold-blooded. But one must not, of course, confuse the superficial appearance of lack of feeling with the actual exercise of professional skill and efficiency. The animal experimenter, likewise, must be professional and effective. My own experience is that animal work makes the investigator *more* sensitive to animal needs, as he learns about their behaviour, their physiology, and gains experience in seemingly small but important matters about how to handle them. One might indeed wonder how it comes about that those who are trying to discover more about the working of the animal body, often explicitly with a view to improving human or animal welfare, should be thought to be *less* sensitive to animal needs than the common run of humanity.

Is animal experiment adequately supervised?

It has been contended that because there appear to have been no prosecutions under the 1876 Act, the inspectors must be failing to provide the control required, and that in effect experiment is uncontrolled. Historically the statement about prosecution is not correct.[8] There have been, so far as I am aware, three prosecutions, which are worth a brief summary. The first was in 1876, when a certain Dr Arbrath was prosecuted for advertising a public lecture on poisons in which (unspecified) experiments would be shown. The advertisement went to press before, and appeared three days after, the passing of the Act. Although in the end no experiment was performed, Dr Arbrath was convicted with a nominal fine. It is to be noted that he belonged to the local branch of the SPCA (as it then was), and that they refused to prosecute. A second case was in 1881, with a prosecution of Dr David Ferrier by the Victoria Street Society (precursor of the National Anti-vivisection Society), for performing experiments on the brain while unlicensed and uncertificated for such experiments. The prosecution failed because the operations involved were in fact performed by Dr G. F. Yeo, who held the required licence and certificates. Ferrier is famous for his work on cortical localization, fundamental to modern neurosurgery, and a Royal Society lecture is named after him. A third case was in 1913. This was a prosecution by the RSPCA of Dr Warrington Yorke for cruelty to a donkey. It involved an experiment in which a drug possibly useful against sleeping sickness produced a type of paralysis. The prosecution failed because Dr Yorke was properly licensed and the suffering involved was not judged to be unnecessary. Dr Yorke, later a Fellow of the Royal Society, was a pioneer of tropical medicine and with Kinghorn discovered the insect vector of one of the sleeping sickness parasites (*Trypanosoma rhodesiense*), an essential piece of knowledge for its control. One further case is of historical interest, since it probably contributed significantly to the movement which resulted in the 1876 Act. This concerned Dr Magnan, a French investigator who demonstrated at a public meeting of the British Medical Association in 1874 that while injection of alcohol into a dog produced anaesthesia, the injection of absinthe produced convulsions. He did this to draw attention to the dangers of absinthe, now well recognized. The RSPCA prosecuted under Martin's Act (the first act in this country dealing with cruelty to animals), but Magnan had by then left the country.

In fact, prosecutions are rather beside the point. There is an immediate sanction available, namely withdrawal of licence or certificates from an individual, or from the premises. For the vast majority of scientists who do animal experiments, such a withdrawal would radically affect their careers, and could make some lose their jobs. No data appear to be available on how often this has been done, but it is evidently extremely rare. This might be interpreted either as a sign of the ineffectiveness of the Inspectorate in missing offences, or of their effectiveness in preventing them. The latter seems more probable in the light of the Annual Reports, which record each year their detection of a series of infringements.

In the great majority of cases, these are technical, such as allowing a licence or certificate to run out while experiments continue. The essential point to which the inspectors' attention is directed is whether any unnecessary suffering was caused by the infringement. Appropriate warnings are given and in two recent cases, the facts were brought to the attention of the Director of Public Prosecutions. Infringements sometimes arise from work by visitors to this country from other countries with less stringent regulations.

It is also contended that adequate supervision cannot possibly be conducted by only fifteen inspectors. This implies some misunderstandings. Firstly, experiments take place in a relatively limited number of 'registered places'—518 at the end of 1982. To these, the Inspectorate made 6,531 visits, mainly without notice: that is an average of 12.6 visits a year to each centre. There were 11,800 active licensees, an average of 23 at each centre. The inspector already knows the nature and purpose of the experiments to be done, and has usually discussed them with the licensee, whom he knows personally. It must be appreciated that the inspector is not a policeman, and that it would be as impracticable as it would be futile to attempt to oversee every experiment. The procedure is, in fact, to get to know the scientists concerned, and to act so as to *prevent* infringements. It is of interest to compare all this with any other inspectable activity. The last available figures for inspection of factories, where considerable human hazard may exist, were of the order of one a year at each establishment. The reader will be able to think of a number of other situations where pain and suffering by humans or animals occur, where inspection of any sort falls considerably short of these figures. We may well conclude that the irony about animal experiment again reveals itself: of all activities which may cause suffering, only that which offers the prospect of less suffering in the future is heavily inspected.

The question of public assurance and public responsibility

Much has been said about the 'secrecy' of animal experiment. This arises primarily from the fact that only the Inspectorate has effective right of access to experimental work. It is therefore thought that there is no information available to the public about the nature of the experiments conducted beyond what is in the Home Office annual reports. This ignores the fact of the vast body of biomedical literature, in which the whole range of experimental procedures, objectives, and results is available in libraries. What is published will not, of course, give an account of every experiment, since in the nature of research some experiments yield results not in themselves worth publishing, however much they lead on to later work. The Littlewood Report mentioned a comment by two of their witnesses (both Edinburgh surgeons) that they estimated that only one-quarter of work done was published. This has actually been cited as a representative figure, although it is evidently a trivial sample of information. But there seems no reason to doubt that even if the number of experiments published falls substantially short of the total done, nevertheless the general *nature* of the experiments will be fully represented, simply because communication of results is the point of scientific work. There is, therefore, a mass of information available to the public.

The information is, in fact, sometimes surveyed by members of animal-welfare groups. The essential point that emerges when this is done is not improper behaviour by the experimenters, but apparent deficiencies in the law in what it allows. That must always be a matter of personal opinion, but it is instructive to consider some of the cases selected in the field of safety evaluation. These cases were regarded as almost self-evidently revealing an unjustified infliction of suffering.[9] The purpose of the following brief description is to give the reader a chance of making his own assessment.

First was a series of experiments injecting nicotine into monkeys, in whom brain electrodes had been placed to record a characteristic change of electrical activity (from slow waves to fast waves) that appears in the transition from sleepiness to alertness —the so-called 'arousal response'. It was found that the change produced by nicotine resembled natural arousal more closely than does that produced by caffeine or amphetamine (the active substance in 'pep pills'). The significance of this is in relation to the long-standing puzzle of *why* people smoke, which is probably linked with the ability of nicotine to improve vigilance in boring tasks. (It is a truism in other fields that social control is impossible until root causes are understood.) It also raises the question whether it is or is not justifiable to use animals to try to understand, and mitigate, the effects of a widespread but voluntary human practice. A similar question arose earlier over the testing of cosmetics. Is society to provide only for the prudent? Or should we recognize that sooner or later we are all imprudent, and provide accordingly?

Second were tests on a new building material for purposes of insulation, *vermiculite*. Asbestos, for which it was hoped vermiculite would be a satisfactory substitute, can produce cancer in those engaged in its production or use. It seems probable that this is in some way related to the fact that the asbestos fibre is exceptionally thin, considerably smaller in diameter than a single cell. Vermiculite also owes its insulating capacity to its very small particle size, although it has a different physical structure; there was therefore good reason to test it. The test was done by injecting either a sample of vermiculite or a sample of asbestos, matched for weight and particle size, into the pleural cavity (the space between the lungs and the ribs) in rats. After two years the two groups, together with some controls injected with saline, were killed and the lungs examined. All three groups remained apparently in perfect health and put on weight normally. Those treated with asbestos showed the early stages of cancer, but it was still far short of causing any symptoms: this was an important 'positive control', proving that the test would in fact reveal a carcinogenic activity if it was present. Those treated with vermiculite or with saline showed no abnormality. While at first sight it might seem obviously wrong to inject a building material into a rat's lungs, a knowledge of the reasons for the experiment sheds a different light.

Third were some experiments on the arsenical war gas Lewisite and protection against it by British Anti-Lewisite (BAL), already mentioned earlier (p. 120). BAL, discovered at Oxford in the last war, was the first effective antidote to be discovered. As it turned out, Lewisite was not used in World War II, although its forerunner in chemical warfare, mustard gas, had been used in World War I: some people believe that this was because an effective antidote was found. After the war the experiments done during the war were published, in which the capacity of BAL to protect against even really severe eye damage was shown: this required the production of eye damage in control rabbits, and it is these experiments that have often been referred to. It would have been possible not to have done these experiments, but to have waited until human cases arose and then, in uncontrolled conditions, to have tried to find out how best to use BAL, and the length of time after exposure over which it was still effective. But it is doubtful whether many people would have preferred this. BAL in fact proved useful during the war in dealing with the toxic effects of another arsenic-containing substance, neoarsphenamine, used in treating syphilis at that time. Subsequently BAL entered the pharmacopoeia since it was found to be an antidote to acute or chronic poisoning with a range of other heavy metals as well as arsenic. The biochemical idea behind its action was very fruitful, and it was the forerunner of other drugs known as 'chelating agents' with a wide range of uses.

Fourth was a proposed fire extinguisher fluid (bromochlorodifluoromethane). It is one of a general chemical class, useful for extinguishing fires because they are themselves non-inflammable and they and their vapour help to exclude oxygen from the burning material. The group as a whole, however, can exert a number of other actions, which include anaesthesia (such as chloroform), liver damage (such as carbon tetrachloride), or convulsions (such as hexafluoroethyl ether). So it was necessary to test whether the new substance was liable to anaesthetize, convulse, or damage the livers of those exposed to it (firemen, or others nearby). In the event it was found, in the dog, to produce convulsions.

Fifth were experiments with a dye known as crystal violet or gentian violet. This was a well-known antiseptic in use for many years, included in the official National Formulary, and used by doctors in a concentration of 0.5 per cent (5 milligrams per cubic centimetre). Its potentiality in causing eye damage was very poorly documented and not mentioned in standard medical reference works. The experimental work that was criticized showed, amongst other things, that a very small drop containing one-tenth of a milligram could cause long-lasting eye damage. The work was done on the advice of independent clinicians, to define the hazard in terms of onset, duration, nature, and severity of the effect.

Sixth were experiments on pig conditioning. Here electric shocks were used, delivered by a battery-operated 'goad', already widely used in animal husbandry. The shocks were 'aversive' in the sense that the pigs would not willingly accept them in order to obtain food. The situation was similar to an animal confined by an electric fence, unwilling to risk a shock to reach something beyond the fence. Responses to it included defaecation, urination, and squealing; pigs will, however, do all three simply on being transferred from a pen to a lorry. The work was intended to analyse stress in animal husbandry (for example, in transport or in mixing groups of pigs), and to evaluate under controlled conditions tranquillizing drugs that could help to reduce such stress.

Seventh were the 'smoking beagles'. The object of these experiments was to test out a possible substitute for tobacco leaf as a smoking material. The damage done by smoking is well documented, and millions of people smoke. The substances in smoke actually responsible for the damage are not known, although there are a number of likely candidates, including tar, which may well act together. The firm involved had succeeded in developing a material which according to a number of tests (for instance on the tar obtained from the smoke) offered appreciable advantages over tobacco, and it was envisaged for use mixed with tobacco leaf. It was required of them to test the smoke in some way directly comparable with human use. In the experiments, beagles were trained to smoke for limited periods, but could always remove their heads from the inhalation masks used. The plan was not to wait until gross bodily changes occurred but to kill the animals after a suitable time to study the early changes (if any) produced. A variety of responsible independent visitors all testified to the absence of any signs of distress in the animals. The experiment could have produced valuable information, but in the event, in response to public reaction, it was terminated and almost no useful information was obtained. The development of safer smoking materials was also virtually stopped in its tracks for some time. There are many questions that arise: some are technical, such as whether the results of such work on the respiratory tract of the dog could be reliably extrapolated to man; but the principal one is a recurrent issue already alluded to, that of the use of animals for the safety evaluation of substances in widespread voluntary human use.

The last case was that of experiments on visual function in kittens. They involved procedures such as sewing up the eyelids of kittens, and would be generally regarded as unacceptable except for a really good reason. That reason was the existence of a condition of functional blindness of the eye known as 'amblyopia': 'functional' means that there is no obvious physical damage, yet the eye does not see. This can arise as a result of the existence of a 'vulnerable period' in the nervous development during early life (at a few years of age). During this period, interference with vision appears able to prevent proper development of the part of the brain concerned with vision. The purpose of the experiments was to analyse the processes involved. The work has already improved the management of eye operations on children (such as for squint), but has a deeper significance for knowledge of the important adaptive changes going on in the brain during childhood. The experiments involved the most meticulous technique, since assessment of any visual changes required behavioural tests which in turn needed healthy and co-operative animals.

The purpose in citing these cases is to show that the public—and particularly the media—also have a responsibility: namely to make sure, in making its own mind up about the justifiability of animal experiment, that it knows or is told the *reasons* for experiments, the background, and the nature of the benefit to knowledge or use that may result. In all these cases the information required was freely available in the literature.

Summary

This chapter deals with how the various considerations involved in animal work are balanced, as revealed in practice.

1. It is essential that any animal experiments done are of satisfactory standard scientifically. The means of ensuring this are considered by reviewing the criticism to which the scientist is exposed during his work—a critique as severe as that in any walk of life.

2. The trend in animal use, now falling, and the various categories of experimental work are surveyed, using the 1982 Home Office figures. In general, the numbers of the various species used diminish as one moves upwards towards the 'higher' animals (from 2,442,702 mice to 5,654 primates and 475 equidae).

3. In assessing whether the numbers used are large or small, it is necessary both to appreciate into how many different categories the work falls, and the numbers of observations needed to *establish* any particular conclusion (illustrated also by opinion polls).

4. Experiments wholly under anaesthesia (about 4 per cent) involve as little suffering as the putting down of a domestic pet. But experiments without anaesthesia (about 80 per cent) or with recovery from anaesthesia (about 16 per cent) are more common. The former are only allowed when a trivial procedure is used, for which anaesthesia would provide a greater stress; for any other than trivial procedures, anaesthesia is required. The main issue is that of pain or suffering that results subsequently. Both types of experiment are subject to the 'pain condition'.

5. Despite claims to the contrary, experiments applying substances to the eye (0.45 per cent), burning or scalding by any means (0.00004 per cent), on cosmetics (0.4 per cent) or on food additives (0.4 per cent) are limited in extent.

ly one-quarter of the experiments done are required
for legislative reasons, and in another third (selection
dental, or veterinary products) the needs of legislative
nvolved.

7. The history of the damage by cosmetics and of the need for more efficient estimates of toxicity that led to the Draize tests and the LD_{50} respectively are reviewed. For both tests, methods involving less or no suffering are being evolved.

8. To embark on animal experiment an individual requires stipend, laboratory space, and research expenses (all in short supply), and must apply to the Home Office for a licence and certificates with appropriate support. His work is then subject to scrutiny by technicians, scientific colleagues, learned societies, journals, and grant-giving bodies. Quite apart from his own feelings of humanity, these factors are a powerful constraint on animal use.

9. The inspectoral system is reviewed. The principal sanction is not prosecution but withdrawal of licence from an individual or from premises—both essential to scientific employment. Animal experiment may claim to be perhaps the most inspected human activity.

10. The public also has a responsibility to ensure that before taking a decision about the justifiability of particular experiments, it understands the scientific context, purpose, and use.

11

THE FUTURE

It has been the theme of this book that there are no 'knock-down' arguments, but rather that animal experimenter and layman alike must balance evidence and arguments over a wide field in order to arrive at accurate and responsible judgement.

It is hoped that the summaries at the end of each chapter will help readers to bring all the points together in their own minds. Rather than recapitulate them, we may now turn to future developments.

There are a good many hopeful signs. Britain has led the way for a hundred years in the care of experimental animals. The issues concerned have recently been debated with great thoroughness.[1] The European Convention at last brings hope that comparable standards will be established elsewhere.[2] If some principles of special importance for legislation must be named, my own would be:

(a) that the numbers of experiments done must be returned; this has given a realism to discussion of practice in Britain lacking elsewhere, and provides an invaluable index of trends;
(b) the inspectoral principle offers very considerable and flexible advantages, allows a cadre of independent expertise to be built up, and facilitates uniformity of standards;
(c) responsibility for the conduct of an experiment must remain with the experimenter himself, so that it is the person most closely in contact with the animal that is responsible for its well-being;
(d) at present, work not directed to discovery but to production, for instance of vaccines, is not included under the Act: this should be remedied, together with the provision of generally agreed codes of conduct in animal care. In this the contribution of the veterinary surgeon and the animal technician must be taken into account;
(e) however legislation is shaped, it must be broadly drawn, so as to be adaptable to changing circumstances.

Further, there are many directions in which animal suffering is being and will be progressively reduced. In toxicity testing, ways are being found to reduce the number of tests needed. Both in toxicity testing and in animal experiment generally, advances in technique allow the use of methods which are either 'non-invasive' (in other words, need no surgery) or require less operative interference. More sensitive methods also allow the earlier detection of structural or functional changes. Our knowledge of methods of relieving pain and suffering has greatly improved, and its application in experimental work offers appreciable scope. The standards of scientific work are rising, increasing the amount of knowledge gained from each experiment. There is now a substantial modern literature on animal anaesthesia,[3] on the care and management of laboratory animals,[4] and on the various alternative methods available[5]. Boundaries between the disciplines are becoming less watertight, allowing useful exchange of techniques and ideas.

Behind it all lies the question of the pattern of future developments, in animal care, in scientific advance, in legislation, and in benefit. I believe Lord Justice Moulton's dictum remains true:

> Your duty is to take that line which produces the minimum total pain, and whether the pain is inflicted pain, or whether it is preventable pain that is not prevented, is in my opinion one and the same thing.[6]

Today, to 'pain' one would add 'suffering'. Moulton in that passage did not refer to the perpetuation of ignorance. We would also admit today that there is no simple utilitarian calculus by which we can 'minimize' pain, but that we simply balance all the considerations as best we can. But the objective he stated remains our task, involving the assessment of any inflicted suffering on the one hand, and on the other the assessment of preventable suffering, as judged from the historical record and the suffering we see around us. Animal experimenters must play their part; and earlier chapters have outlined the nature of the work they do, the non-animal alternatives they use, the context in which they work, and the critique and inspection to which they are subject. Russell and Burch in 1959 advanced a programme called the '3 Rs' concerning the use of animals in experimental work: 'replacement' by alternatives; 'reduction' in numbers; and 'refinement' of method.[7] But there is a fourth R, 'responsibility'; and it is one that the world outside the laboratory also shares.

Has it been responsible for the media to have focused so often only on experimentally inflicted animal suffering, and virtually to have ignored both the historical record of animal and human suffering relieved and all the preventable suffering waiting for relief? Has it been responsible for the animal hooligans, mostly of the healthiest generation this country has ever seen because of past medical research, to harass the experimenters, the animal breeders, and the industries? Has it been responsible for individuals to say that while, of course, they cannot condone violence, yet if something is not done soon one must not be surprised at 'direct action'—can condonation go further in a peaceful democracy? Animal experiment is essential in medical and veterinary research. Those who unreasonably harass and restrict it now, will carry a grave responsibility for the unnecessary ignorance and unnecessary human and animal suffering that will result in the future.

NOTES

Full details of the works referred to here are given in the Bibliography (pp. 163–8).

Chapter 1

1 Marshall Hall 1831, pp. 2–8
2 Macaulay 1876; Philanthropos 1883; Paget 1900; Smith 1901; Coleridge 1906; Westacott 1949; French 1975; Moss 1961
3 Walder 1983
4 Ryder 1975, p.2
5 Brock 1975
6 Rogers 1937
7 Lepage 1960

Chapter 2

1 Orwell 1945
2 Leewenhoek (in Dobell) 1932
3 Schweitzer 1933
4 Eisely 1959
5 Kavaliers *et al.* 1983
6 Darwin 1875, p. 462
7 French 1975
8 Cruelty to Animals Act 1876
9 Littlewood 1965
10 Mountcastle 1968, Ch. 63
11 *European Convention* 1983
12 Moss 1961

Chapter 3

1 Dale 1953
2 Rothschild 1971
3 Dainton 1971

Chapter 4

1 Royal Commission on Vivisection 1907–12; Macaulay 1876; Westacott 1949
2 Bentham 1789, Ch. 17, paragraph 4
3 Singer 1976. One may distinguish two arguments in Singer's book: (1) that animal suffering deserves equal consideration with that in man, because both are sentient; (2) that 'speciesism', treating man and animals differently because they are different species, is comparable to sexism and racism. But surely we do not believe that men and women, and black and white, should be treated equally just because they are all sentient, but because they can participate fully in the whole range of human activity.
4 Caplan 1983
5 Frey 1983
6 Hull 1976
7 Linzey 1976, p. 22
8 Ritchie 1916, p. 81
9 Clark 1982, p. 124
10 Gladstone 1896
11 Dunstan 1979; Halsbury 1973

Chapter 5

1 Kennedy 1968, p. 272
2 Garrison 1924; Majno 1975
3 North 1983, p. 82
4 Paris 1822
5 D'Arcy Hart 1946
6 Office of Health Economics 1966
7 Anderson 1977
8 Paton 1976a
9 Medical Research Council 1977
10 Billsborough 1983
11 Pollock and Morris 1983
12 *British Medical Journal* 1983
13 Anderson and May 1982
14 Rogers 1937
15 Lehner *et al.* 1980
16 Paton *et al.* 1978
17 Burchenal and Krakoff 1956
18 Shadwell 1911
19 Doll and Peto 1981
20 Adapted from Doll and Peto 1981, Table D1
21 Doll 1983, p. 93; Table adapted from Table IV
22 Learmonth 1954; Adams and Bell 1977
23 Wiles and Devas
24 Comroe and Dripps 1977; Lapage 1960; Barrett 1955
25 Brent 1965; Woodruff 1972
26 Beeson 1980

27 Osler 1909
28 Browne 1600, Ch. 5
29 Weatherall 1982
30 Geddes 1981
31 *British Pharmacopoeia 1982* and *British Pharmacopoeia (Veterinary) 1977*
32 Herriot 1972 and 1974
33 Ewald and Gregg 1983
34 Goodwin 1980
35 Shaper 1972
36 Hull *et al.* 1983
37 *British Medical Journal* 1975
38 Lainson 1982
39 Moulton 1907–1912: the first quotation is from paragraph 12790, the second from 12737

Chapter 6

1 Comroe and Dripps 1977
2 Doll 1983

Chapter 7

1 Lewis 1942
2 Dawkins 1980
3 Medawar 1972, p. 80: see also 'On the use of animals in medical research', Medical and Health Annual, *Encyclopaedia Britannica* (Chicago, 1983)

Chapter 8

1 Littlewood 1965, paragraphs 248–56 and General Finding (6)
2 Bawden and Brock 1982
3 Smyth 1978
4 Holmstedt and Liljestrand 1963
5 Campbell *et al.* 1983
6 Foster and Langley 1876
7 Home Office 1983, Table 21
8 Medical Research Council *Annual Reports* 1963–82
9 Gowans 1974
10 Paton 1979b
11 But see Jefferson 1955 and Lepage 1960
12 Pappworth 1967: see also Elliott 1974
13 Garrison 1924
14 Holmstedt and Liljestrand 1963: Serturner's experiment is described on p. 74, Purkinje's on p. 87, and Christison's on p. 95
15 Smith, S. M., *et al.* 1947
16 Head 1920, pp. 225–329; Trotter and Davies 1909
17 Lewis 1927
18 Lewis 1942
19 Keele and Armstrong 1963
20 Mountcastle 1968
21 Medical Research Council 1977

Chapter 9

1 Spinks 1963
2 Lester and Keoskosky 1967
3 Flovey *et al.* 1949

Chapter 10

1 US House of Representatives 1952
2 US Committee on Labour and Public Welfare 1976
3 Draize *et al.* 1944
4 Department of Prices and Consumer Protection 1976
5 Trevan 1927
7 British Toxicology Society Report 1984
8 *Conquest* 1977
9 *Conquest* 1976

Chapter 11

1 House of Lords 1980
2 *European Convention* 1983
3 Green 1982
4 UFAW 1967
5 Turner 1983; Sechzer 1983; Williams *et al.* 1983
6 Moulton 1907–1912, paragraph 12704
7 Russell and Burch 1959

BIBLIOGRAPHY

Adams, I. W., and Bell, M. S. (1977), 'A comparative trial of polyglycolic acid and silk as suture materials for accidental wounds', *Lancet*, **ii**, 1216–17.

Altman, D. G. (1980), 'Statistics and ethics in medical research: VI Presentation of results', *British Medical Journal*, **281**, 1542–4.

Anderson, R. M., and May, R. M. (1982), 'Directly transmitted infectious diseases: control by vaccination', *Science*, **215**, 1053–60.

Anderson, T. (1977), 'The role of medicine', *Lancet*, **i**, 747.

Barrett, N. R., (ed.) (1955), 'Surgery of the heart and thoracic blood vessels', *British Medical Bulletin*, **11**, 171–242.

Bawden, D., and Brock, Alison M. (1982), 'Chemical toxicology searching: a collaborative evaluation, comparing information resources and searching techniques', *Journal of Information Science*, **5**, 3–18.

Beeson, P. B. (1980a), 'How to foster the gain of knowledge about disease', *Perspectives in Biology and Medicine*, Winter 1980, Part 2.

Beeson, P. B. (1980b), 'Changes in medical therapy during the past half century', *Medicine*, **59**, 79–99.

Bentham, J. (1789), *An Introduction to the principles of morals and legislation*.

Billsborough, J. S. (1983), 'Whooping cough: the public health viewpoint', *Health Trends*, **15** (3), 71–3.

Brent, L. (1965), 'Transplantation of tissues and organs', *British Medical Bulletin*, **21**, 97–182.

British Medical Journal (1975), 'Immunisation in the two-thirds world' (editorial), **v**, 369.

British Medical Journal (1983), 'Views', **i**, 1288.

British Pharmacopoeia 1982, (HMSO, London).

British Pharmacopoeia (Veterinary) 1977 (HMSO, London).

British Toxicology Society Report (1984), *Human Toxicology*, **3** (2), 85–92.

Brock, Lord (1975), Hansard, 14 May, cols. 758–9; *The Times*, 8 May and 23 October 1981; *Times Higher Educational Supplement*, 5 February 1982.

Browne, T. (1643), *Religio Medici*.

Browne, T. (1658), *Hydriotaphia, Urne Buriall, or, a Discourse of the Sepulchrall Urnes lately found in Norfolk*

Burchenal, J. H., and Krakoff, I. H. (1956), 'Newer agents in the treatment of leukemia', *Archives of Internal Medicine*, **98**, 567–73.

Campbell, W. C., Fisher, M. H., Stapley, E. O., Albers-Schöhberg, G., and Jacob, T. A. (1983), 'Ivermectin: a potent new antiparasitic agent', *Science*, **221**, 823–8.

Caplan, A. J. (1983), 'Beastly Conduct: Ethical Issues in Animal Experimentation', *Annals of the New York Academy of Sciences*, **406**, 159–69.

Clark, S. R. L. (1982), *The Nature of the Beast: Are Animals Moral?* (Oxford University Press, Oxford and New York).

Coleridge, S. (1906), *Vivisection: a Heartless Science* (John Lane, The Bodley Head, London).
Comroe, J. H., and Dripps, R. D. (1977), *The top ten clinical advances in cardiovascular-pulmonary medicine and surgery 1945–1975* (US Government Printing Office, Washington DC).
Conquest (1976) No. 167, 'Some notes on experiments recently criticized'.
Conquest (1977) No. 168, 'Prosecutions under the 1876 Act'.
Cruelty to Animals Act (1876), 39 and 40 Vict. Ch. 77.
Dainton, Sir Frederick (1971), 'The Future of the Research Council System', *A Framework for Government Research and Development*, Cmnd 4814 (HMSO, London).
Dale, H. H. (1953), *Adventures in Physiology* (Pergamon Press, London).
Darwin, C. (1875), *Insectivorous Plants* (John Murray, London).
Dawkins, Marion S. (1980), *Animal Suffering: the Science of Animal Welfare* (Chapman and Hall, London).
Department of Prices and Consumer Protection (1976), *Report by Consumer Safety Unit on Childhood Poisoning from Household Products* (Consumer Safety Unit, DPCP, London).
Doll, R. (1983), 'Cancer control', The Lilly Lecture, *Symposium on Medical Management of Malignant Disease* (Royal College of Physicians, Edinburgh).
Doll, R., and Peto, R. (1981), *The Causes of Cancer* (Oxford University Press, Oxford and New York).
Draize, J. H., Woodard, E., and Calvery, H. O. (1944), 'Methods for the study of irritation and toxicity of substances applied topically to the skin', *Journal of Pharmacology and Experimental Therapeutics*, **82**, 377–90.
Dunstan, Canon G. R. (1979), 'A limited dominion', Paget Lecture, *Conquest*, No. 170, pp. 1–8.
Eisely, Loren (1959), *Darwin's Century* (Gollancz, London).
Elliott, A. H. (1974), *Medical Experimentation* (Medical Research Council of New Zealand, PO Box 6063, Dunedin).
European Convention for the Protection of Vertebrate Animals used for Experimental and Other Scientific Purposes 1983 (Council of Europe, Strasbourg).
Ewald, B. H., and Gregg, D. A. (1983), 'Animal research for animals', *Annals of the New York Academy of Sciences*, **406**, 48–58.
Florey, H. W., Chain, E., Heatley, N. G., Jennings, M. A., Sanders, A. G., Abraham, E. P., and Florey, M. E. (1949), *Antibiotics* (Oxford University Press, Oxford), Vol. II, Part 8.
Foster, M., and Langley, J. N. (1876), *A Course of Elementary Practical Physiology* (Macmillan, London).
French, R. (1975), *Antivivisection and Medical Science in Victorian Society* (Princeton University Press, Princeton and London).
Frey, R. G. (1983), 'Vivisection, morals and medicine', *Journal of Medical Ethics*, **9**, 94–7.
Gaddum, J. H. (1933), 'Methods of Biological Assay depending on a Quantal Response', *Medical Research Council Special Report*, Series No. 183 (HMSO, London).
Garrison, F. H. (1924), *History of Medicine* (Saunders, Philadelphia and London).
Geddes, A. M. (1981), 'Infection in Britain today', *Journal of the Royal College of Physicians, London*, **15**, 100.

Gladstone, W. E. (1896), *Studies Subsidiary to Butler's Works* (Clarendon Press, Oxford).

Goodwin, L. G. (1980), 'New drugs for old diseases', *Transactions of the Royal Society of Tropical Medicine and Hygiene*, **74**, 1–7.

Gowans, J. L. (1974), 'Alternative methods to animal experiment in medical research', Paget Lecture, *Conquest*, No. 165, pp. 2–6.

Green, C. J. (1982), *Animal Anaesthesia*, 2nd edition (Laboratory Animals Ltd., London).

Hall, Marshall (1831), 'A Critical and Experimental Essay on the Circulation of the Blood' (Sherwood, Gilbert, and Piper, London): reprinted in 1847 in *Lancet*, **i**, 58–60, 135, 161.

Halsbury, Lord (1973), 'Ethics and the exploitation of animals', Paget Lecture, *Conquest*, No. 164, pp. 2–12.

Hart, P. D. (1946), 'Chemotherapy of tuberculosis', *British Medical Journal*, **ii**, 805.

Head, H. (1920), *Studies in Neurology* (Oxford University Press, London) Vol. I.

Herriot, J. (1972), *It Shouldn't Happen to a Vet* (Michael Joseph, London).

Herriot, J. (1974), *Vet in Harness* (Michael Joseph, London).

Holmstedt, B., and Liljestrand, G. (1963) *Readings in Pharmacology* (Pergamon Press, Oxford and London).

Home Office (1983), *Statistics of Experiments on Living Animals, Great Britain 1982*, Cmnd 8986 (HMSO, London).

House of Lords Select Committee on the Laboratory Animals Protection Bill (1980), Vol. II, *minutes of evidence* (HMSO, London).

Hull, D. L. (1976), 'The rights of animals', *Science*, **192**, 679–80; review of *Animal Liberation* by P. Singer.

Hull, H. H., Williams, P. J., and Oldfield, F. (1983), 'Measles mortality and vaccine efficacy in rural West Africa', *Lancet*, **i**, 972.

Jefferson, Sir Geoffrey (1955), 'Man as an experimental animal', Paget Lecture, *Conquest*, No. 141, pp. 2–11.

Kavaliers, M., Hirst, M., Teskey, G. C. (1983), 'A functional role for an opiate system in snail thermal behaviour', *Science*, **220**, 99–101.

Keele, C. A., and Armstrong, Desiree (1963), *Substances Producing Pain and Itch*, Physiological Society Monograph No. 12 (Edward Arnold, London).

Kennedy, M. (1968), *Portrait of Elgar* (Oxford University Press, Oxford).

Lainson, R. (1982), 'Leishmanial parasites of mammals in relation to human disease', *Symposium of the Zoological Society, London*, **50**, 137–79.

Learmonth, Sir James (1954), 'The surgeon's debt to animal experiment', *Conquest*, No. 137, pp. 3–10.

Leeuwenhoek, A. v., in Dobell, C. (1932), *Antony van Leeuwenhoek and his Little Animals* (Staples Press, London).

Lehner, T., Russell, M. W., and Caldwell, J. (1980), 'Immunisation with a purified protein from *Streptococcus mutans* against dental caries in Rhesus monkeys' *Lancet*, **i**, 995–6.

Lepage, G. (1960), *Achievement. Some Contributions of Animal Experiment to the Conquest of Disease* (W. Heffer and Sons Ltd., Cambridge).

Lester. D. and Keoskosky, W. Z. (1967), *Life Sciences*, **6**, 2313–19.

Lewis, T. (1927), *The Blood Vessels of the Human Skin and their Responses* (Shaw and Sons, London).

Lewis, T. (1942), *Pain* (Macmillan, New York).

Linzey, A. (1976), *Animal Rights* (SCM Press, London).

Littlewood, Sir Sydney (1965), Report of the Departmental Committee on Experiments on Animals. Cmnd 2641 (HMSO, London).

Macaulay, J., (no date given, c.1876), *Plea for Mercy to Animals* (Religious Tract Society, London).

Majno, G. M. (1975), *The Healing Hand: Man and Wound in the Ancient World* (Harvard University Press, Cambridge, Mass.).

Medawar, P. B. (1972), *The Hope of Progress* (Methuen, London).

Medical Research Council (1947), 'Medical Research in War', *Report of the Medical Research Council for the years 1939–1945*, Cmnd 7335 (HMSO, London).

Medical Research Council (1963–82), *Annual Reports* (HMSO, London).

Medical Research Council (1977), 'Clinical trial of live measles vaccine given alone and live vaccine preceded by killed vaccine', Report by Measles Subcommittee, *Lancet*, **ii**, 571–5.

Moss, A. W. (1961), *Valiant Crusade: The History of the R.S.P.C.A.* (Cassell, London).

Moulton, Lord Justice, in *Royal Commission on Vivisection* (1907–1912), paras. 12691–818.

Mountcastle, V. B. (1968), *Medical Physiology* (C. V. Mosby Company, St. Louis).

North, R. (1983), *The Animals Report* (Penguin, Harmondsworth).

Office of Health Economics (1966), *Disorders which Shorten Life*, Report No. 21 (Office of Health Economics, London).

Orwell, G. (1945), *Animal Farm* (Secker and Warburg, London).

Osler, W. (1909), 'The treatment of disease', *British Medical Journal*, **ii**, 185–9.

Paget, S. (1900), *Experiments on Animals*, with an Introduction by Lord Lister (T. Fisher Unwin, London).

Pappworth, M. H. (1967), *Human Guinea-pigs: Experimentation on Man* (Routledge and Kegan Paul, London).

Paris, J. A. (1822), *Pharmacologia*, 5th edition (W. Phillips, London).

Paton, W. D. M. (1979a), 'Animal experiment and medical research: a study in evolution', Paget Lecture, *Conquest*, No. 169, 1–14.

Paton, W. D. M. (1979b), 'The evolution of therapeutics: Osler's therapeutic nihilism and the changing pharmacopoeia', Osler Oration Lecture, *Journal of the Royal College of Physicians*, **13**, 74–83.

Paton, W. D. M., Zaimis, Eleanor, Black, J. W., and Green, A. F. (1978), 'High Blood Pressure: The Evolution of Drug Treatment: British Contribution', in *Highlights of British Science* (The Royal Society, London).

Philanthropos (1883), *Physiological Cruelty: or Fact v. Fancy* (Tinsley and Co., London).

Pollard, R. (1983), 'Whooping cough in Fiji', *Lancet*, **i**, 1381.

Pollock, T. M., and Morris, Jean (1983), 'A 7-year survey of disorders attributed to vaccination in North West Thames Region', *Lancet*, **i**, 753–7.

Ritchie, D. G. (1916), *Natural Rights*, 3rd edition (Allen and Unwin, London).

Rogers, Sir Leonard (1937), *the Truth about Vivisection* (Churchill, London).

Rothschild, Lord (1971), 'The organization and management of government research and development', in *A Framework for Government Research and Development* Cmnd. 4814 (HMSO, London).

Royal Commission on Vivisection, Reports I–VI and Final Report, 1907–1912 (HMSO, London).

Russell, W. M. S., and Burch, R. L. (1959), *The Principles of Human Experimental Technique* (Methuen, London).

Ryder, R. D. (1974), *Scientific Cruelty for Commercial Profit* (Scottish Society for the Prevention of Vivisection).

Sather, H., Miller, D., Nesbit, M., Heyn, Ruth, and Hammond, D. (1981), 'Differences in prognosis for boys and girls with acute lymphoblastic leukemia', *Lancet*, **i**, 739.

Schweitzer, A. (1933), *My Life and Thought: an Autobiography* (Allen and Unwin, London).

Sechzer, Jeri A. (1983), 'The role of animals in biomedical research', *Annals of the New York Academy of Sciences*, **406**.

Shadwell, A. (1911), 'Cancer' in *Encyclopaedia Britannica*, 11th edition (Cambridge University Press, Cambridge).

Shaper, A. G. (1972), 'Cardiovascular disease in the tropics. I. Rheumatic heart', *British Medical Journal*, **iii**, 683–6.

Singer, P. (1976), *Animal Liberation* (Jonathan Cape, London).

Smith, S.(no date given, *c.* 1901), *Scientific Research: a View from Within* (Elliott Stock, London).

Smith, S. M., Brown, H. O., Toman, J. E. P., and Goodman, L. S. (1947), 'Lack of cerebral effects of d-tubocurarine' *Anaesthesiology*, **8**, 1–14.

Smyth, D. H. (1978), *Alternatives to Animal Experiment* (Scolar Press, London).

Spinks, A. (1963), 'Justification of clinical trial of new drugs', *Proceedings of the Second International Pharmacological Meeting* (Prague), **8**, pp. 7–19.

Trevan, J. W. (1927), 'The error of determination of toxicity', *Proceedings of the Royal Society*, B, **101**, 483–514.

Trotter, W., and Davies, H. M. (1909), 'Experimental studies on the innervation of the skin', *Journal of Physiology*, **38**, 134.

Turner, P. (ed.) (1983), *Animals in Scientific Research: an Effective Substitute for Man?* (Macmillan, London).

UFAW Handbook (1967), *The Care and Management of Laboratory Animals*, 3rd edition (Livingstone, Edinburgh and London).

US House of Representatives (1952), Hearings by Select Committee to Investigate the Use of Chemicals in Foods and Cosmetics (US Government Printing Office, Washington DC).

US Committee on Labor and Public Welfare, Cosmetic Safety Amendments 1976, Hearing before the Subcommittee on Health (US Government Printing Office, Washington DC).

Walder, A. (1983), quoted in *The Times*, 22 April.

Weatherall, Josephine S. (1982), 'A Review of some effects of recent medical practices in reducing the numbers of children born with congenital abnormalities', *Health Trends*, **14** (4), 85–8.

Westacott, E. (1949), *A Century of Vivisection and Antivivisection* (C. W. Daniel Company Ltd., Ashington, England).

Wiles, P., and Devas, M. B. (1954), 'The halt and the maimed', *Conquest*, No. 138, pp. 2–6.

Williams, G. M., Dunkel, V. C., and Ray, V. A. (eds.) (1983), 'Cellular systems for toxicity testing', *Annals of the New York Academy of Sciences*, **407**, pp. 1–484.

Woodruff, Sir Michael (1972), 'The contribution of animal experiments to the surgery of replacement', *Conquest*, No. 163, pp. 3–7.

INDEX

MORE OXFORD PAPERBACKS

The Nature of the Beast

Are Animals Moral?

Stephen R. L. Clark

Many of us would brand uncouth behaviour as 'animal-like'. Yet we half believe that some animals live in peace and friendship, that our own sad history is an aberration, and that we can learn from the beasts.

Stephen Clark attempts to come to grips with these contradictory tendencies, and to find a rational reconciliation between them. He discusses the evidence for animal intelligence, and whether concepts such as 'freedom', 'self' and 'obligation' can be applied to animals. He also analyses sexual behaviour, parenting and dominance, and argues that human moralizing rests on sentiments that we experience because we are mammals.

'clear, lively, convincing. A remarkable achievement.' Mary Midgley, *New Scientist*

The Nature of Human Aggression

Ashley Montagu

Is man a born killer? In one of the most important books of his career, Ashley Montagu debunks this currently fashionable theory. He takes issue with the innate aggressionists Konrad Lorenz, Robert Ardrey, Niko Tinbergen, Desmond Morris, and others — and shows that 'on every one of the fundamental claims they have made concerning man's allegedly instinctive aggressive drives, they are demonstrably wrong'.

' a vigorous book attacking the nonsense . . . concerning the inherent aggressiveness of Man. It should not still need writing.' *Guardian*

'a counsel of hope rather than one of despair – and one based on a realistic, cogently-argued view of man's nature' *Times Educational Supplement*

The Reproduction Revolution

New Ways of Making Babies

Peter Singer and Deane Wells

On 25 July 1978 in Oldham, Lancashire, Louise Brown was born. She was the first test-tube baby, and with her began a new era in childbirth. Now, five years after this event, *in vitro* fertilization and other technological advances are bringing hope to infertile couples in clinics throughout the world.

The authors of this accessible new book argue that it is high time we faced up to the complex ethical problems of the reproduction revolution. They explore the new processes through the eyes of both the doctor, and of the potential parents. They also look to the future when ectogenesis (development outside the womb), cloning, sex selection, and genetic engineering may become commonplace.

Pluto's Republic

Incorporating The Art of the Soluble
and Induction and Intuition in Scientific Thought

Sir Peter Medawar

What *is* science? What sort of person is a scientist? What kind of reasoning leads to scientific discovery? These are the central questions to which Sir Peter constantly returns in these wide-ranging studies. The answers are often surprising.

'The 1960 winner of the Nobel Prize for Medicine sheds light on popular misconceptions and *ideés fixes* in the field of scientific thought. Valuable explanations of topics we assume we understand but would be hard pressed to clarify.' *Sunday Telegraph*

'A definitive edition of Medawar's essays' Bernard Dixon, *New Scientist*

J. B. S.

The Life and Work of J. B. S. Haldane

Ronald Clark

Preface by *Sir Peter Medawar*

'An excellent biography . . . both just and warm-hearted.' C. P. Snow, *Sunday Times*

This is the definitive biography of one of the most brilliant of British scientists. Haldane was a trail-blazing geneticist and physiologist, a highly successful science populariser, a dedicated Marxist, and a devotee of Hindu culture whose tempestuous private and public life made him one of the most controversial figures in the scientific world.